# Doubling Farmers' Income through Agroforestry

**Abhay Kumar**

**Assistant Professor (Agroforestry)**

Department of Agronomy,

Ranchi Agriculture College, Birsa Agricultural University,

Ranchi, Jharkhand, India

**Swati Shabnam**

**Assistant Professor (Agronomy)**

Department of Agronomy

Ranchi Agriculture College, Birsa Agricultural University

Ranchi, Jharkhand, India

**M. S. Malik**

**Chairman and University Professor**

Department of Silviculture and Agroforestry

Faculty of Forestry, Birsa Agricultural University

Ranchi, Jharkhand, India

SCIENTIFIC PUBLISHERS

Published by :

**SCIENTIFIC PUBLISHERS (INDIA)**
5 A, New Pali Road, P.O. Box 91
Jodhpur 342 001 (India)
E-mail: info? scientificpub.com
Website: www.scientificpub.com



ISBN : 978-93-90495-71-9
eISBN : 978-93-90495-74-0

Printed in India

**Dr. S. Bhaskar**
Assistant Director General (ADG)
(Agronomy, Agroforestry and Climate Change)
Indian Council of Agricultural Research (ICAR)
KAB-II, PUSA, New Delhi-110012
Email-*adgagroandaf@gmail.com*

# Foreword

The Government of India in 2016 announced to double the farmers' income (DFI) by 2022 by shifting the focus from agricultural output and food security to income security through diversification of agriculture. Agroforestry being the integration of tree, crop, animal/pasture in the same land at the same time, is one of the suitable options for enhancing overall farm income, as income from a single crop or commodity will not fulfill the target of doubling farmers' income.

Agroforestry is a land use system that integrates trees and shrubs on farmlands and rural landscapes to enhance productivity, profitability, diversity and ecosystem sustainability. Agroforestry is the key path to prosperity for farmers and rural people, thus leading to the generation of employment and revenue; food and nutritional security; meeting the other basic human needs on the sustainable basis and cushioning farmers from the harshness of climate change. Agroforestry creates more integrated, diverse, productive, profitable, healthy, and sustainable land use systems and is the only option to increase the country's forest and tree cover to 33 percent.

India is the first country in the world to adapt a National Agroforestry Policy in the year 2014. Implementation of National Agroforestry Policy, Sub-Mission on Agroforestry, National Bamboo Mission, National Horticulture Mission and various other policies and schemes clearly demonstrate the commitment and will of the Government of India to mainstream agroforestry at large scale.

I am therefore happy that the present book on "**Doubling Farmers' Income through Agroforestry**" written by Abhay Kumar, Swati Shabnam and M. S. Malik has brought together a summary of the works done in this field. We owe a deep debt of gratitude to the authors for their hard work and love for the science which will help to enhance farmers' income through agroforestry.

I firmly believe that this publication will be highly useful, in one way or the other, for researchers, academicians, extension workers, policy makers, planners, officials in development institutions/agencies, producers, farmers and students of agroforestry, forestry and agriculture.

**(S. Bhaskar)**

# Preface

This book deals with importance of opportunities for doubling farmers' income, enhancing farmer' income, employment generation, improving food and nutritional security, enrichment of soil, air and water quality, enhancing biodiversity, aesthetic and recreational value, carbon sequestration and climate change and mitigation through agroforestry.

Agroforestry systems have a potential to produce 100 million cubic meter timber/pulpwood for industrial and domestic use, which fulfills 65 percent of the country's timber demand, 2/3 of small timber demand, 70-80 percent of plywood demand, 60 percent raw material demand for paper pulp, 9-11 percent of the green fodder demand and also produces 150 million tons of firewood.

This book contains 20 chapters naming introduction, policy & scheme, IFS, sericulture, lac culture, apiculture and economically important tree species (Teak, Gamhar, Poplar, Malabar neem, Eucalyptus, Sandalwood and Bamboo), horticulture/fruit tree species (Mango, Aonla, Guava, Coconut, Cashew nut, and Medicinal and Aromatic plants) based agroforestry system which gives higher gross income, net return, B:C ratio than open farming system (sole crops and agroforestry system).

Keeping in mind these points, we wrote the book on "Doubling Farmers' Income through Agroforestry" and suggest farmers for cultivation of different high income, important tree species based agroforestry system which will help in income generation or doubling farmers' income in India.

— **Authors**

# Acknowledgement

I express my sincere gratitude to Dr. M. S. Yadava, Dean, Faculty of Agriculture, Dr. M. S. Siddaqui, Dean, Faculty of Forestry, Dr. P. R. Oraon, Assistant Registrar, Faculty of Forestry, Birsa Agricultural University, Ranchi, Jharkhand, Dr. Virendra Singh, University Professor, Dr. Salil Tiwari, Director of Students' Welfares, GBPUAT, Pantnagar, Uttarakhand, Dr. Suman Jha, Dr. Satish Sinha, Dr. N. S. Thakur, College of Forestry, Navsari Agricultural University, Gujarat and Dr. Ashok Kumar, PAU, Ludhiana, Panjab for their valuable suggestions and support while writing this book. I also feel highly indebted to my respected seniors & lovely friends Dr. Vikas Rena, JNU, New Delhi, Dr. Ram Gopal, FRI, Dehradun, Indra Singh, VCSG-UUHF, Uttarakhand, Bijay Kumar Singh, and Ankit Kumar Jha BAU, Ranchi, India, for their constant support and help in the documentation of research and writing this manuscript. I am also grateful to all persons involved directly or indirectly in preparation of this work.

## About Authors

**Abhay Kumar** is currently working as an Assistant Professor (Agroforesry) in Department of Agronomy, Ranchi Agriculture College, Birsa Agricultural University, Ranchi, Jharkhand. He has completed B.Sc. (Hons) Forestry from NAU, Gujarat and M.Sc. Agroforestry from GBPUAT, Pantnagar, Uttarakhand and Ph.D. Forestry (Pursuing), specialization in Agroforestry, BAU, Ranchi, Jharkhand. He has also qualified NET three times from ASRB-ICAR, New Delhi and UGC-NET in Environmental science. He has been awarded with Indo Global Excellence Award, Young Teacher Award and Young Scientist Award-2018 and also got NTS from July 2009 to June 2013 by ICAR. He has participated in and completed more than 10 online MOOC courses conducted by IIRS, NAARM, and Elsevier etc. He is life member of many societies and editorial member of journals. He has got more than 15 research papers, 3 book chapters and a number of articles in magazine/bulletin published and more than 20 seminar/ symposia presented or attended. He possesses 3.5 years of teaching and research experience in the areas of agroforestry, forestry, environment science, agriculture etc.

**Swati Shabnam** is currently working as an Assistant Professor (Agronomy) in Department of Agronomy, Ranchi Agriculture College, Birsa Agricultural University, Ranchi, Jharkhand. She has completed B.Sc. (Hons) Agriculture from SHUATS, Prayagraj, U.P., and M.Sc. and Ph.D. Agronomy from Birsa Agricultural University, Ranchi. She has qualified NET from ASRB-ICAR, New Delhi. She has published many research papers, 2 book chapters, magazine/bulletin and participated/ presented paper in many national/international seminar and symposia.

**Dr. M. S. Malik** holds a Doctoral degree in Forestry from Tamil Nadu Agricultural University, Coimbatore. He possesses 32 years of teaching and research experience in the areas of agroforestry, poverty alleviation and participatory natural resource management. At present, he is working with Birsa Agriculture University as Professor and Chairman, Department of Silvicultur and Agroforesty and also Principal Investigator of GOI-World Bank Sponsored National Agricultural Higher Education project. His overall research experience encompasses agroforestry, Forest management, community-based natural resource management and poverty alleviation. He has published several papers on related issues.

# Contents

# Index of Tables

# Index of Figures

## List of Symbols used

| | |
|---|---|
| @ | At the rate of |
| a.i | Active ingredient |
| b | Billion |
| B:C | Benefit cost ratio |
| BHC | Benzene Hexa Chloride |
| C.D. | Critical difference |
| Cft | Cubic feet |
| cm | Centimeter |
| cum | Cubic meter |
| CV | Coefficient of variation |
| *et al.* | Coworkers and others |
| °C | Degree Celsius |
| DAS | Days after sowing |
| d.f. | Degree of freedom |
| EC | Emulsifiable Concentrate |
| EC | Electrical Conductivity |
| etc. | Et. cetera |
| f.b. | Followed by |
| Fig. | Figure |
| FYM | Farm Yard Manure |
| g | Gram |
| ha | Hectare |
| h | Hour |
| IBA | Indole-3 Butyric Acid |
| Kcal | Kilo calorie |
| kg | Kilogram |
| km | Kilometre |
| L | Litre |
| m | Metre |
| m ha | Million Hectare |
| mm | Millimetre |
| min | Minute |
| ml | Millilitre |
| MT | Million tons |
| MOP | Muriate of Potash |
| N | Nitrogen |
| NPK | Nitrogen, Phosphorus and Potassium |
| No. | Number |
| pH | Negative logarithm of the reciprocal of the $H^+$ ion activity |
| NAA | Naphthalene Acetic Acid |
| % | Percent |
| $P_2O_5$ | Phosphorus |
| ppm | Parts per million |
| $K_2O$ | Potash |
| q | Quintal |
| RBD | Randomised Block Design |
| R.H. | Relative humidity |
| RDF | Recommended Dose of Fertilizer |
| Rs. | Rupees |
| S | Second |
| SSP | Single Super Phosphate |
| Spp. | Species |
| $m^2$ | Square metre |
| Temp. | Temperature |
| *viz.* | That is to say/ in others words |
| t | Tons |
| i.e. | That is |
| Yr | Year |

## List of Abbreviations used

AICRP – All India Coordinated Research Project
ALUS – Alternate Land-Use Systems
APMC – Agricultural Produce Market Committee
ARDC – Agricultural Refinance and Development Corporation
AYUSH – Ayurveda, Yoga & Naturopathy, Unani, Siddha & Homoeopathy
BAIF – Bharatiya Agro Industries Foundation
BAU – Birsa Agricultural University
BCR – Benefit-Cost Ratio
CAFRI – Central Agroforestry Research Institute
CAMPA – Compensatory Afforestation Fund Management and Planning Authority
CDB – Coconut Development Board
CGIAR – Consultative Group for International Agricultural Research
CO2 – Carbon Dioxide
COP – Conference of the Parties
CPTs – Candidate Plus Trees
DFI – Doubling Farmer's Income
eNAM – National Agriculture Marketing
ETPs – Entire Transplants
FAO – Food and Agriculture Organization
FAR – Global Forest Resources Assessment
FCA – Forest (Conservation) Act
FRI – Forest Research Institute
GDP – Gross Domestic Product
GHGs – Green House Gases
GIM – Green India Mission
GIS – Geographic Information System
GOI – Government of India
ICAR – Indian Council of Agricultural Research
ICFRE – Indian Council of Forestry Research and Education
ICRAF – International Council for Research in Agroforestry
IFGTB – Institute of Forest Genetics and Tree Breeding

IFS – Integrated Farming System
INDC – Intended Nationally Determined Contribution
IRR – Internal Rate of Return
ISFR – India State of Forest Report
ITC – Imperial Tobacco Company
IWST – Institute of Wood Science and Technology
KAU – Kerala Agricultural University
KFRI – Kerala Forest Research Institute
KSDL – Karnataka Soaps and Detergent Limited
KVK – Krishi Vigyan Kendra
LEV – Land Expectation Value
MAPs – Medicinal and Aromatic Plants
MIDH – Mission for Integrated Development of Horticulture
MLO – Mycoplasma Like Organisms
MoEF&CC – Ministry of Environment, Forests and Climate Change
NABARD – National Bank for Agriculture and Rural Development
NAEB – National Afforestation and Eco-Development
NAFCC – National Adaptation Fund for Climate Change
NAP – National Agriculture Policy
NAP – National Agroforestry Policy
NAPCC – National Action Plan on Climate change
NBM – National Bamboo Mission
NFP – National Forest Policy
NGOs – Non Government Organization
NHM – National Horticulture Mission
NITI – National Institution for Transforming India
NMPB – National Medicinal Plants Board
NMSA – National Mission for Sustainable Agriculture
NPV – Net Present Value
NRCAF – National Research Centre for Agroforestry
PBP – Payback Period
PKVY – Paramparagat Krishi Vikas Yojana
PMFBY – Pradhan Mantri Fasal Bima Yojana
R & D – Research and Development
RBI – Reserve Bank of India

| | | |
|---|---|---|
| REDD | – | Reducing Emissions from Deforestation and Forest Degradation |
| RFRI | – | Rain Forest Research Institute |
| SAPCC | – | State Action Plan on Climate Change |
| SDGs | – | Sustainable Development Goals |
| SMAF | – | Sub-Mission on Agroforestry |
| TFRI | – | Tropical Forest Research Institute |
| TNAU | – | Tamil Nadu Agricultural University |
| TP | – | Transit Permit |
| UNFCCC | – | United Nations Framework Conservation on Climate Change |
| UN | – | United Nations |
| USDA | – | United States Development Agency |
| WAFC | – | World Agroforestry Centre |
| WIMCO | – | Western India Match Company |
| WTO | – | World Trade Organization |

## CHAPTER - 1

# Introduction to Agroforestry

### Introduction

Agroforestry is a collective name for land-use systems and technologies where woody perennials (trees, shrubs, palms, bamboos, etc.) are deliberately used on the same land-management units as agricultural crops and/or animals, in some form of spatial arrangement or temporal sequence. In an agroforestry system there are both ecological and economical interactions between the various components (Lundgren and Raintree, 1982).

**This definition implies that:**

- Agroforestry normally involves two or more species of plants (or plants and animals), at least one of which is a woody perennial.
- An agroforestry system always has two or more outputs.
- The cycle of an agroforestry system is always more than one year.
- Even the simplest agroforestry system is more complex, ecologically (structurally and functionally) and economically, than a mono-cropping system

Agroforestry is an integration of trees and herbaceous crops for economic and ecological benefits (Rajalingam *et al.,* 2016). It is a sustainable land management system which increases the overall yield of the lands; combine the trees and shrubs with agricultural crops and or livestock on the same unit of land, either simultaneously or sequentially. One of the main problems that farmers face today is decreasing income from a hectate per year against sudden increase in the value of agricultural lands. In this context, combination of agricultural crops with trees ensures profitability, productivity and sustainable utility (Thapa *et al.,* 1989). Intercropping pastures/fodder crops with tree species is one of the sustainable practices in agroforestry particularly for fodder production.

Agroforestry helps in the sustainability of agriculture and its diversification to attain higher benefits per unit area of land, when carefully selected and managed with the introduction of fast growing multipurpose trees with little effect on agricultural crops (Johl *et al.,* 1986).

## Importance of Agroforestry

1. Enhances suitable farm productivity
2. Improves food and nutrition security
3. Helps reduce poverty
4. Contributes to create resilient livelihood
5. Contributes to climate change mitigation and adaptation
6. Solves social issues
7. Enriches soil
8. Improves air and water quality
9. Improves biodiversity
10. Aids industrial growth
11. Provides employment opportunity
12. Enhances aesthetic value of area

## Benefits from Agroforestry can be grouped under

### Environmental benefits-

1. Reduction of pressure on forest.
2. Reduction of surface run-off, nutrient leaching and soil erosion.
3. More efficient recycling of nutrients by deep rooted trees on the site.
4. Better protection of ecological systems.
5. Improvement of microclimate.
6. Increment in soil nutrients through addition and decomposition of litter fall.
7. Improvement of soil structure through the constant addition of organic matter from decomposed litter.

### Economic benefits

1. Increment in maintenance of outputs of food, fuel-wood, fodder, fertilizer and timber
2. Reduction in incidence of total crop failure, common to single cropping or monoculture system
3. Increase in levels of farm income.

### Social benefits

1. Improvement in rural living standards from sustained employment and higher incomes.
2. Improvement in nutrition and health due to increased quality and diversity of food outputs.

## National benefits

1. Sustained supply of raw materials to forest based industries.
2. Reduction in hunger and poverty by increasing production of farms for both consumption and sale.
3. Reduction in the speed of climate change by more carbon sequestration and storage by trees.

Agroforestry plays a vital role in the Indian economy by way of tangible and intangible benefits. In fact, agroforestry has high potential for simultaneously satisfying three important objectives *viz.*, protecting and stabilizing the ecosystems; producing a high level of output of economic goods and improving income and basic materials to rural population. Agroforestry is a land use system, which integrates trees and shrubs on farmlands and rural landscapes to enhance productivity, profitability, diversity and ecosystem sustainability. Agroforestry not only provides environmental services, but also economic gains, as about 65 percent of the country's timber requirement is met from the trees grown outside forests (SMAF, 2016). Wood based products such as paper, pulp and furniture offer vast potential for enhanced income for farmers (NITI, 2017). At present agroforestry meets almost half of the demand of fuel wood, two third of small timber, 70–80 percent wood for plywood, 60 percent of raw material for paper pulp and 9–11 percent of the green fodder requirement of livestock, besides meeting the subsistence needs of households for food, fruit, fibre, medicine, timber etc. (ICAR-CAFRI, 2015).

Agroforestry provides both climate change mitigation and adaptation benefits to farmers. In addition to income from sale of agroforestry products, trees on farms are critical component of climate smart agriculture in many systems. Trees regulate soil moisture, moderating drought or heavy precipitation and soil temperature. Trees contribute to soil fertility by adding nutrients in the case of nitrogen fixing species and they contribute to increasing soil organic matter. Increasing carbon biomass on agricultural lands through agroforestry may also improve biodiversity, water quality and in some cases, hydrological cycles (CGIAR, 2017). Moreover, in 2016, India ratified the Paris Agreement and committed itself to reducing the emission intensity of its gross domestic product (GDP) upto 33–35 per cent from 2005 level; to increase the share of non-fossil based power generation to 40 per cent; and to create an additional carbon sink of 2.5 to 3.0 billion tons of carbon dioxide ($CO_2$) equivalent through additional forest and tree cover by 2030 under Intended Nationally Determined Contribution (INDC) submitted to United Nations Framework Conservation on Climate Change (UNFCCC). Likewise, the United Nations (UN) Sustainable Development Goals (SDGs) emphasize climate, forests and bio-energy as pillars to achieve all 17 goals by 2030. Given the nature of multi-functionality of agroforestry, it also has roles to play in achieving the targets of SDG1 (on ending poverty), SDG2 (on ending hunger), SDG6 (on protecting water related ecosystems), SDG7 (on access to sustainable energy to all), SDG13 (on combating climate change) and SDG15 (on protect terrestrial ecosystems). This is a current way of sustainable land management practiced in India and its contribution to food, fuel, fodder, fiber, fertilizer, timber, climate change, carbon sequestration, employment generation is well recognized.

In this way it is also recognized that agroforestry is perhaps the only easy alternative to meeting the target of increasing forest cover to 33 per cent from the present level of less than 25 per cent, as envisaged in the National Forest Policy (1988). According to Planning Commission report "Greening India", 33% forest cover can only be achieved through agroforestry. Prior to the 'Green Revolution', the majority of subsistence farming anywhere in the world involved growing of mixed species, usually including trees (Mark and Mbow, 2014).

Agriculture and forestry can no longer be treated in isolation and linking these two is imperative for socio-economic development in the twenty-first century (FAO, 2016). In the documented agroforestry practices, trees serve as wind breaks and shelterbelts, delineate boundaries, and provide shade, ornamentation and seclusion around homesteads. They supply not only poles, stakes, timber and fuel, but also cash crops, fodder, fruits and nuts, dyes, gums, resins, fiber and medicines. Fodder and food trees provide balanced diets during dry seasons, when other foods are scarce.

## History of agroforestry

Agroforestry is a new name for the old practice of growing trees on farmland. Agroforestry was formally outlined in the early 20th century by J. Russell Smith in his book Tree Crops: A Permanent Agriculture (1929). Smith viewed tree-based "permanent agriculture" as a solution to the destructive erosion that often followed the cultivation of sloping lands. However, his contributions were largely overlooked during the green revolution of the 1960s and the subsequent and more inclusive farming systems research or extension development approach of the early 1970s that sought sustainable agricultural alternatives. The term 'agroforestry' was first coined in 1977 to describe the integration of trees and agriculture. International Council for Research in Agroforestry (ICRAF) was established in 1977 with its headquarter in Nairobi, Kenya and Dr. Kenneth King was appointed first Director General. In 2002 ICRAF was renamed as World Agroforestry Centre (WAFC). In India, National Research Centre for Agroforestry (NRCAF) Jhansi, Uttar Pradesh was established in 1988. NRCAF was renamed as Central Agroforestry Research Institute (CAFRI) in 2014. All India Coordinated Research Project (AICRP) on Agroforestry was started in 1983 with 20 centers in the country. At present there are 37 centers of AICRP on Agroforestry located in 27 State Agricultural Universities (SAUs), 09 in ICAR and 01 in ICFRE Institutes representing all agroclimatic zones of the country.

## Area under agroforestry

World has a total forest area of 4.06 billion hectares (ha), which is 31 percent of the total land area. This area is equivalent to 0.52 ha per person, although forests are not distributed equally among the world's peoples or geographically. Major top ten countries in descending order are Russia, Brazil, Canada, USA, China, Congo, Australia, Indonesia, Peru and India. Ninety-three percent (3.75 billion ha) of the forest area worldwide is composed of naturally regenerating forests and 7 percent (290 million ha) is planted. The area of naturally regenerating forests has decreased since 1990 (at a declining rate of loss), but the area of planted forests has increased by 123 million ha (FAO, 2020). It means the agroforestry area increased. As per the World Bank 2004 report, an estimated 1.2 billion rural people currently practice agroforestry on their farms and in their communities and depend upon its products. Almost 10% of the world's agricultural area is under agroforestry, whereas Nair *et al.* 2009, estimated the total area under agroforestry was 2.35 billion ha.

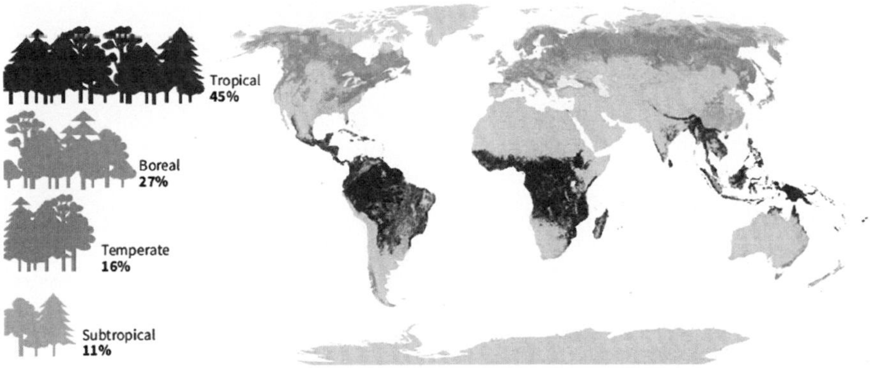


**Fig.1.1:** World forest map 2020 (FAO, 2020).

In India, according to India State of Forest report (ISFR, 2019), the total forest and tree cover is 80.7276 million hectare, which is 24.56% of the total geographical area. The estimated total green cover in the agroforestry system of the country is estimated as 111,554 sq km which is 3.39 percent of country's geographical area (ISFR, 2013). Among all the states, Uttar Pradesh, Maharashtra and Rajasthan is ranked first, second and third in terms of area under agroforestry (1.86, 1.61 and 1.55 m ha, respectively). Dhyani *et al.* 2014, the area under agroforestry is estimated to be 25.32 m ha, or 8.2% of the total geographical area of the country. Based on data from CAFRI, Jhansi and Bhuvan LISS III, the area under agroforestry is 13.75 m ha (Rizvi *et al.* 2016). The estimation by Forest Survey of India, Dehradun does not include many agroforestry practices such as block plantations, thus reflecting lower values than CAFRI. The differences in the approximate figures by the two different organizations necessitated uniform and accurate assessment methods based on remote sensing and GIS application. Further, fine-tuning of area for a particular agroforestry system like agrisiliviculture, silvipasture, agrisilvihorti-culure, traditional systems, area under a particular tree-based system is also to be looked into and it will be a challenging task for research organizations. This can be achieved by developing digital signatures for all the important agroforestry tree species. Table-1.1, shows the total area under agroforestry in different agro-climatic zones of India.

**Table 1.1:** Agro-climatic zone wise estimated agroforestry area in India.

| ACR No. | Agro-climatic zones | Total Geographical Area (M ha) | Area under Agroforestry (M ha) | Agroforestry Area (%) |
|---|---|---|---|---|
| 3 | Lower Gangetic Plains Region | 6.733 | 0.802 | 11.91 |
| 4 | Middle Gangetic Plains Region | 16.570 | 1.304 | 7.87 |
| 5 | Upper Gangetic Plains Region | 14.441 | 2.234 | 15.47 |
| 6 | Trans Gangetic Plains Region | 11.603 | 1.143 | 9.85 |
| 8 | Central Plateau & Hill Region | 37.843 | 37.843 | 5.09 |
| 9 | Western Plateau & Hill Region | 32.740 | 1.556 | 4.75 |
| 10 | Southern Plateau & Hill Region | 39.412 | 2.976 | 7.55 |
| 12 | West Coast Plains & Hill Region | 11.682 | 1.657 | 14.18 |
| 13 | Gujarat Plains & Hill Region | 18.977 | 2.570 | 13.56 |
| 14 | Western Dry Region | 14 17.871 | 0.431 | 2.41 |
| **Total** | | **207.872** | **16.599** | **7.98** |

*Source-* Ram Newaj *et al.*, 2017, CAFRI

## Development of Agroforestry in India

For development and promote agroforestry, different types of organization like Government of India, state governments, policy makers, NGOs etc. Agroforestry was incorporated into national agricultural and forestry research agendas when ICAR launched an All India Coordinated Research Project (AICRP) on Agroforestry with 20 centers in 1983 followed by the establishment of Central Agroforestry Research Institute (CAFRI) on 08 May 1988 at Jhansi, U.P. At present there are 37 centers of AICRP on Agroforestry located in 27 State Agricultural Universities (SAUs), 09 in ICAR and 01 in ICFRE Institutes representing all agro-climates of the country. In addition to ICAR, Indian Council of Forestry Research and Education (ICFRE) also conducts agroforestry research through its research institutes and advanced research centers in different parts of the country (ICFRE 2011). Recognizing agroforestry as a viable venture, many business corporations, limited companies such as ITC, WIMCO, West Coast Paper Mills Ltd., Hindustan Paper Mills Ltd., and other institutions initiated agroforestry research with emphasis on production of improved planting material of the fast growing species

(Dhyani *et al.*, 2015)

Forest (Conservation) Act-1980 (FCA) reoriented India's forestry policy towards conservation. The FCA specified that the "felling of trees in private plantations shall be governed by various State Acts and Rules". The National Forest Policy, which aimed at bringing 33 per cent of the country's area under forest cover, further cemented the view that trees should not be cut.

For development and promotion of agroforestry, Ministry of Environment, Forests and climate change (MoEF&CC), Government of India issued new guidelines in 2014 that advised states to exempt popular agroforestry species from transit permit (TP), if these trees are not naturally occurring in the forest. All over India from 2017, bamboo grown on agricultural land will not require a transit permit (TP), for felling and transportation, from the forest department.

**Table 1.2**: List of exempted and restricted tree species from different states of India

| State | Exempted (no permission) | Restricted species (permission required) |
|---|---|---|
| Andhra Pradesh | *Mangifera indica, Casuarina equisetifolia, Acacia nilotica, Syzygium cumini, Psidium guajava, Azadirachta indica, Anacardium occidentale, Cocus nucifera, Ficus religiosa, Eucalyptus* spp. and *Leucaena leucocephala* | *Pterocarpus santalinus, Tectona grandis* and *Santalum album* |
| Assam | *Aquilaria malaccensis, Santalum album, Mangifera indica, Syzygium cumini, Artocarpus integrifolia, Eucalyptus tereticornis, Poplar deltoides, Psidium guajava etc.* | — |
| Bihar | *Poplar deltoides, Eucalyptus tereticornis, Anthocephalus cadamba, Mangifera indica, Bombax ceiba, Gmelina arborea, Litchi chinensis, Phoenix sylvestris* and *Bamboo* spp. | *Tectona grandis* |
| Chhattisgarh | *Many* spp. | *Tectona grandis, Madhuca latifolia, Dalbergia latifolia, Tamarindus indica* and *Shorea robusta* |
| Gujarat | *Eucalyptus tereticornis, Casuarina equisetifolia* and *Prosopis juliflora* | *Tectona grandis, Mangifera indica, Madhuca latifolia, Acacia catechu, Dalbergia latifolia, Santalum album,* |

| State | Exempted (no permission) | Restricted species (permission required) |
|---|---|---|
| | | *Madhuca indica, Gmelina arborea* and *Anogeissus pendula* |
| Himachal Pradesh | 23 species permitted. Some of the important ones are *Albizia procera, Bauhinia racemosa, Eucalyptus* spp., *Morus alba, Populus deltoides, Saussurea costus, Bunian persicum, Broussonetia papyrifera, Celtis orientalis, Toona ciliata, Syzygiun cumini, Terminilia arjuna, Mallotus philippensis* and *Bombax ceiba* | — |
| Jharkhand | *Leucaena leucocephala, Bamboo* spp., Cane, *Acacia nilotica, Shorea robusta, Tectona grandis, Pterocarpus marsupium, Boswellia serrata* and *Terminalia elliptica* | *Mangifera indica, Madhuca indica, Gmelina arborea, Dalbergia sissoo, Artocarpus heterophyllus* and *Syzygium cumini* |
| Kerala | 61 species permitted. Some of the important ones are *Pongamia pinnata, Hevea brasiliensis, Ailanthus excelsa, Mangifera indica, Garcinia cambogia* and *Terminalia* spp. | *Santalum album, Tectona grandis, Dalbergia latifolia, Xylia xylocarpa, Hopea parviflora, Michelia champaca, Grewia tiliifolia* and *Toona ciliate* |
| Karnataka | 42 species permitted. Some of the important ones are *Acacia hybrid, Acacia mangium, Alianthus excels, Semenea saman, Araucaria columnaris, Erythrina variegate, Citrus sinensis, Citrus limon, Areca catechu, Anacardium occidentale Casuarina equisetifolia, Coffea arabica Cocos nucifera, Epigaea repens Eucalyptus* spp. etc. | *Dalbergia latifolia, Santalum album* and *Acacia catechu* |
| Madhya Pradesh | *Eucalyptus* spp., *Casuarina equisetifolia Leucaena leucocephala, Populus deltoides, Pithocellobium dulce* and *Prosopis juliflora* | *Acacia nilotica, Albizia lebbeck, Azadirachta indica, Zizyphus mauritiana, Butea monosperma, Syzygium cumini, Acacia leucophloea* and *Bamboo* spp. |

| State | Exempted (no permission) | Restricted species (permission required) |
|---|---|---|
| Maharashtra | *Acacia nilotica, Leucaena leucocephala, Acacia auriculiformis, Prosopis juliflora, Eucalyptus* spp., *Moringa olefera, Manilkara zapota, Populus deltoides* and *Phoenix sylvestris* | *Terminalia chebula, Tectona grandis, Madhuca latifolia, Tamarindus indica, Mangifera indica, Artocarpus integrifolia, Acacia catechu, Santalum album, Pterocarpus marsupium, Adina cordifolia, Terminalia tomentosa, Hardwickia binata Syzigium cumini* and *Mangroves* spp. |
| Meghalaya | 37 species permitted. Some of the important ones are *Populus* sp., *Leucaena leucocephala, Acacia auriculiformis, Acacia nilotica, Ailanthus excelsa, Anthocephalus cadamba, Aquilaria maiaccensis, Delonix regia, Jaccaranda musifolia, Grevillea robusta* etc. | — |
| Mizoram | 21 species permitted. Some of the important ones are *Bauhinia varigata, Bauhinia pupurea, Erythrina variegate, Hevea brasiliensis, Sterculia urens, Grevillea robusta, Eucalyptus* spp. etc. | — |
| Odisha | *Bambusa nutan, B. vulgaris, B. tulda, Samania saman, Eucalyptus hybrid, Acacia auriculiformis, Cassia siamea, Casuarina equisetifolia* and *Grevillea robusta* | *Mangifera indica, Artocarpus heterophyllus, Schleichera oleosa, Madhuca indica, Tamarindus indica* and *Santalum album* |
| Punjab and Haryana | *Eucalyptus tereticornis, Populus deltoides Bamboo* spp., *Morus alba, Psidium guajava Alianthus excelsa* and *Melia composite* | *Acacia catechu, Emblica officinalis* and *Bamboo* spp, |
| Rajasthan | 39 species permitted. Some of the important ones are *Eucalyptus* spp., *Leucaena leucocephala, Alianthus excelsa, Acacia tortilis, Morus alba,, Putranjiva roxburghii, Psidium* | |

| State | Exempted (no permission) | Restricted species (permission required) |
|---|---|---|
| | *guajava, Syzygium cumini* and *Tamarindus indica* | |
| Tamil Nadu | *Grevillea robusta, Eucalyptus* spp., *Casuarina equisetifolia* and *Leucaena leucocephala* | *Santalum album, Diospyros melanoxylon, Pterocarpus santalinus, Dalbergia latifolia* and *Tectona grandis* |
| Telangana | 42 species permitted. Some of the important ones are *Eucalyptus* spp., *Cocos nucifera, Pinus roxburghii, Leucaena leucocephala, Jacaranda mimosifolia, Morus alba, Anthocephalus cadamba, Anacardium occidentale, Borassus flabellifer Gmelina arborea, Alstonia scholaris, Samanea saman* etc. | *Tectona grandis, Dalbergia latifolia, Santalum album, Pterocarpus santalinus* etc. |
| Uttarakhand | 23 species permitted. Some of the important ones are *Populus deltoides, Eucalyptus* spp., *Leucaena leucocephala,, Grevillea robusta, Pinus* sp., *Salix* sp., *Cupressus* sp. etc. | *Populous ciliate, Pinus roxberghii, Pinus wallichiana, Cupressus torulosa, Acacia catechu, Cordyceps* sp., etc. |
| Uttar Pradesh | *Eucalyptus tereticornis, Populus deltoids, Leucaena leucocephala, Casuarina equisetifolia, Ailanthus* spp., *Gmelina arborea, Grevillea robusta, Morus alba, Anthocephalus cadamba, Melia composita, Acacia* spp., *Albizia* spp., *Borassus flabeliformis, Butea monosperma, Tamarindus indica* and *Grewia oppositifolia* | *Mangifera indica, Madhuca indica, Ficus bengalensis, Azadirachta indica* and *Carpinus viminae* |
| West Bengal | No tree exempted | *Acacia catechu, Bombax ceiba, Dalbergia sissoo, Diospyrus melanoxylon, Gmelina arborea, Madhuca indica, Michelia champaca, Shorearo busta, Swietenia mahagony, Tetona grandis* and *Mangroves* spp. |

*Source*- Chavan *et al.*, 2015

## Recent Research Trends in Agroforestry

At present scenario in all over world, agroforestry research and development is very broad area cover like enhance farm economics, employment generation, decrease poverty, zero hunger, food security, climate change and mitigation, control GHGs emission, increase greenery, new evergreen revolution etc.

The recent research trends in agroforestry are as follows:

1. Climate smart agroforestry.
2. Mitigating climate change with agroforestry.
3. Agroforestry for combating land degradation and desertification in dry areas.
4. Agroforestry and biodiversity conservation.
5. Agroforestry for water quality and watershed restoration.
6. Agroforestry and the migration challenge.
7. Agroforestry encourages jobs and green economy.
8. Scaling up of scientifically confirmed agroforestry innovations.
9. Agroforestry and certification.
10. Agroforestry in public policies.
11. Economics of agroforestry: the link between nature and society.
12. Agroforestry value chains.
13. Socio-economic and cultural controls on technology adoption in agroforestry systems.
14. Agroforestry landscape.
15. Urban agroforestry for food and nutritional security.
16. Agroforestry mapping.
17. Agroforestry and business / private sector / finance.
18. Challenges, opportunities and trade-offs of agroforestry with perennial crops (coffee, cocoa, coconut, rubber, sandalwood, orchards…).
19. Silvopastoral system: promoting rural development by increasing tree-animal-pasture relationships.
20. The scale, dynamics and multiple impacts of farmer-managed natural regeneration.
21. The renewal of tree pollarding.
22. Ever green agriculture.
23. Agroforestry and agroecology: opportunities and challenges.
24. Tree genetics resources for agroforestry systems.
25. Agroforestry: pests, diseases and weeds.
26. Roots matters in agroforestry.
27. Biophysical modelling of interactions in agroforestry.

## Reference

Chavan SB, Keerthika A, Dhyani SK, Handa AK, Ram Newaj and Rajarajan K 2015 National Agroforestry Policy in India: A low hanging fruit. *Current Science* **108** 1826–1834

Dhyani SK 2014 National Agroforestry Policy-2014 and the need for area estimation under agroforestry. *Current Science* **107** 9–10.

Dhyani SK, Handa AK and Uma 2015 Agroforestry Research in India: Present Status and Future Perspective. In: Dhyani SK, Newaj R, Alam B and Dev I Eds. Organized Agroforestry Research in India: Present Status and Way Forward. Boitech Book Publisher, Delhi.

FAO 2020 Global Forest Resources Assessment 2020– Key findings, Rome. https://doi.org/10.4060/ca8753en

ISFR 2013 India State of Forest Report, Forest Survey of India, (Ministry of Environment & Forests), Dehradun, India, Chapter-7, 71–80 pages.

ISFR 2019 India State of Forest Report, Forest Survey of India, (Ministry of Environment & Forests), Dehradun, India.

Nair PKR, Kumar BM and Nair VD 2009 Agroforestry as a strategy for carbon sequestration. *J. Plant Nutr. Soil Sci.* **172** 10–23.

Rajalingam GV, Divya MP, Prabakaran C and Parthiban KT 2016 Performance of vegetable crops under Ailanthus excelsa based agroforestry system. *Indian Journal of Agroforestry* **18**(1) 16–20.

Ram Newaj, Rizvi RH, Chaturvedi OP, Alam B, Prasad R, Kumar D and Handa AK 2017 A Country Level Assessment of Area under Agroforestry and its Carbon Sequestration Potential. *Technical bulletin* 2/2017. ICAR-Central Agroforestry Research Institute, Jhansi, 15 page.

Rizvi RH, Dhyani SK, Ram Newaj, Karmakar PS and Saxena A 2016 Mapping agroforestry area in India through remote sensing and preliminary estimates. *Indian Farm.* **63**(11) 62–64.

Thapa B, Joshi L and Sherpa SL 1989 Traditional agriculture and resource use under pressure. Scope for agroforestry intervention in the middle hills of Nepal. PAC occasional paper Pakhnbas Agriculture Centre, 138 page.

Chapter - 2

# Doubling Farmer's Income through Agroforestry

The Hon'ble Prime Minister of India, Shri. Narendra Modi, imagined this target on 28 Februrary 2016 while addressing farmers in Bareilly district of Uttar Pradesh to develop the economic condition of the farmers. The aim year (*i.e.* 2022) has more special connotations since that happens to be the 75th year of the country's Independence. The goal is to double the income of farmers by 2022. For the first time, a Prime Minister has set such a target in front of the nation for the benefit of farmers. Under the guidance of the Prime Minister, the Agriculture Ministry is working to complete the target by 2022. As the policy level negotiations of various stakeholders toward identifying farm income issues and affixing strategies have gathered steam all over the country, it is away from doubt that the Hon'ble Prime Minister's vision has made a paradigm shift in our thinking process from focusing just on food security to ensuring income security.

There cannot be anybody questioning the need for doubling farmers' income. But how far it is possible to be achieved within the targeted five years remains the doubtful question. For the target to be realized, it is estimated that an agricultural growth rate of 10.46 per cent is required. In this context, it is important to note that the latest Situational Assessment Survey of NSSO (2014) illuminating nominal rise in farmers' income pan-India between 2002–03 and 2013–14 has been an impressive 11.4 percent but the real income has only grown by 3.5 percent and also substantiate this declaration as they show mean farmers' income in India to be around Rs. 6,250 per month.

**Why double farmers' income?**

- Past policy for development of the agriculture sector in India has focused primarily on raising agricultural output and improving food security.
- The net result has been a 45 percent increase in food production, which has made India not only food self-sufficient at collective level, but also a net food exporting country.
- The plan did not explicitly recognise the need to raise farmers' income and did not mention any direct measure to encourage farmers welfare.

- ➢ The net result has been that farmers income remained low, which is clear from the incidence of poverty among farm households.
- ➢ Doubling real income of farmers till 2022–23 over the base year of 2015–16, requires annual growth of 10.41 per cent in farmer's income.

Different types of strategy for Doubling Farmers' Income by government of India

A. There is however, almost unanimity that the net income of farmers can surely be doubled well within the period of six years. A total of 40 recommendations have been given for increasing incomes of farmers, which can be broadly divided into five major heads:

1. Increasing incomes by improving productivity
2. Water and Agri-Input policies
3. Integrated Farming System
4. Better market price realization
5. Special Policy Measures

**B. Pledge to Double Farmers Income by 2022 Seven-point Strategy**

The Ministry is working sincerely to fulfill Prime Minister's dream. To double the farmers' income, a large number of officials and farmers have also been taking a pledge at events organized by the KVK since 16 August 2017.

Seven-point Strategy

1. **Increase in production**– It is important to improve irrigation efficiency to increase production and hence, government has increased the irrigation budget with the motto 'Per Drop More Crop'. For this purpose, Pradhan Mantri Krishi Sinchai Yojana has been initiated to mitigate drought by ensuring 'water to every farm'.
2. **Effective use of input cost**– The government has introduced Soil Health Card Scheme to inform farmers about nutrient status of their soils after which farmers are going for balanced use of fertilizers and following the recommended dose of fertilizer for their field which has reduced the cost of cultivation and ultimately enhance income.
3. **Reduction of post-harvest losses**– The government is encouraging farmers to use warehouses and avoid distressed sales. To protect produce to get wasted and farmers suffering from loss, the government is emphasizing on storage facilities and has initiated integrated cold chains in rural areas.
4. **Value Addition**– The government is promoting quality through food processing. Pradhan Mantri Kisan Sampada Yojana has been started and Rs. 6,000 crore has been allocated for this project. Under this scheme, food-processing capabilities will be developed by working on the forward and backward linkage of agro processing cluster, which will benefit 20 lakh farmers and create employment opportunities for about 5,00,000.

5. **Reforms in Agriculture Marketing**– e-NAM was launched with three reforms and so far, 455 mandis have been linked to this platform. Online trading has begun on various mandis. The government has finalized and implemented model Agricultural Produce Market Committee (APMC) Act, which includes private market yards and encourages direct marketing. The Government is also working on a Model Act to promote contract farming.
6. **Risk, Security and Assistance**– The government has initiated Pradhan Mantri Fasal Bima Yojana (PMFBY) to reduce the possible risks. The scheme creates a security shield. The lowest rate has been fixed whereas maximum rate is 2% and 1.5% for *Kharif* and *Rabi* crops respectively. The scheme covers the loss to standing crops as well as pre-sowing to post-harvesting losses. 25% of the claim is settled online immediately. The government is providing compensation if at least 33% of the crop is damaged. Compensation amount has also been increased 1.5 times.
7. **Allied Activities- Horticulture, Agroforestry, Integrated Farming, Beekeeping etc.**– National Agroforestry Policy-2014 and Sub-Mission on Agro forestry-2016, has been initiated with an aim to promote inter-cropping. Under this scheme, "Har Med Par Ped" campaign has been initiated to encourage farmers to plant trees on farms.

Integrated Farming System (IFS)-The focus is also on horticulture, agroforestry, livestock, and bee keeping. Practicing integrated farming by including different enterprises in addition to crops will not only increase farmers' income, it will also mitigate the effect of drought, flood, and other natural disasters.

Horticulture-Mission for Integrated Development of Horticulture (MIDH) scheme is playing an important role in doubling the income of farmers. For this, government is providing with better planting materials and improved seed and also focusing on protected cultivation, high-density plantation, rejuvenation, and precision farming.

## C. Recommendation of M. S. Swaminathan's Report

1. **Irrigation**– Promotion of rainwater harvesting which enables farmers to have water supply through sustained and equitable access and also recharge the aquifer ("Million Wells Recharge" programme).
2. **Agricultural productivity**– Substantial increase in public investment in agriculture related infrastructure particularly in irrigation, drainage, land development, water conservation, research development and road connectivity. A national network of advanced soil testing laboratories with micronutrient testing facilities as micronutrient deficiencies are becoming common nowadays.
3. **Credit and insurance**– Expansion of the formal credit system to reach the really poor and needy. Issuing Kisan Credit Cards to women farmers with joint pattas for collateral use. Expansion of crop insurance to cover the entire country and all crops with reduced premiums.

## How agroforestry helps to Double Farmers Income

- Agroforestry, the integration of trees with annual crop cultivation, livestock production and other farm activities, is a series of land management approaches practised by more than 1.2 billion people worldwide. Integration of various components increases farm productivity when they occupy complementary niches and their mutual associations are managed effectively and efficiently (Steffan - Dewenter *et al.*, 2007).
- Increment in maintenance of outputs of food, fuel-wood, fodder, fertilizer and timber.
- Reduction in incidence of total crop failure, common to single cropping or monoculture system.
- Increase in levels of farm income.
- Improvement in rural living standards from sustained employment and higher incomes.
- Improvement in nutrition and health due to increased quality and diversity of food outputs.
- Agroforestry supports food and nutritional security through the direct provision of food, by raising farmers' incomes and providing fuel for cooking, and through various ecosystem services.
- Agroforestry is an important climate - smart agriculture approach.
- Agroforestry in supporting food and nutritional security include policy and market constraints.
- Developments in agroforestry policies are required to reform tree and land tenure for the benefit of small scale farmers, to reform how smallholders obtain agroforestry inputs such as tree planting material so that agroforestry is recognized as an important investment option.
- Research should support tree domestication to improve yields and enhance the complementarity and stability of food production in smallholder agroforestry systems.
- Agroforestry systems provide a variety of products and services that are important locally, nationally and globally.
- Relatively low - input agroforestry options are often favoured by women who are unable to afford high - cost technologies due to severe cash and credit constraints.

## Role of Agroforestry

The agroforestry system is recognized as an important integrated farming practice since time immemorial to fulfill the necessity of food, fodder, fuel wood, fibre and timber along with aesthetic and environmental services. However, woody prennials were gradually reduced from farmlands due to increasing demand of foodgrains to fulfill the needs of increasing population on limited land resources in entire world in general and India in particular. In the wake of climate change and declining factor productivity, all governments are supporting agroforestry system due to its role in soil-health improvement, nutrient cycling, carbon sequestration and better economic return as compared to existing cropping systems with less pressure on natural resources. In 2011, the United States Development Agency (USDA) developed a strategic framework for agroforestry for the period 2011–16 to strengthen agroforestry research and extension. In India, major policy initiatives, including the National Agricultural Policy, 2000; Planning Commission Task Force on Greening India 2001; National Forest Policy 1952, 1988 and 2018(Draft); National Horticulture Mission, 2005; National Bamboo Mission, 2006; Green India Mission, 2010; National Mission for Sustainable Agriculture (NMSA), 2014; National Agroforestry Policy, 2014; Sub-Mission on Agroforestry (SMAF), 2016 etc. emphasized the role of agroforestry for sustained productivity and to achieve the goal of 33% forest cover. Bihar and Chhattisgarh has taken the initiative on agroforestry policy to alleviate poverty and provide an alternate source of income to farmers.

Growing of marketable trees on farms is seen as more profitable option than field crops in many regions of India. However, in spite of relative higher profitability to farmers and several concerted efforts made by government, the adoption of agroforestry has not yet reached the expected desired level. Two common factors determining the adoption of agroforestry in any specific region are relative profitability with other crops, and price stability of wood.

The fast growing exotic tree species like Poplar (*Populus deltoides*), Malabar neem (*Melia azadichta*) and Eucalyptus (*Eucalyptus tereticornis*) have been introduced at farmers' field to get maximum income from a given land in a short period of time. These tree species proved as economically viable option for crop diversification in the region.

## Activities associated with agroforestry and its employment generation potential

Employment growth with rising productivity is the most effective mechanism to alleviate poverty of the poor with equity. Agroforestry provides employment for the unskilled rural sectorwith relatively lower investment (Dhyani *et. al.*, 2003). The role of agroforestry in employment generation through various activities involved in production and processing is well documented. Studies reveal that agroforestry systems have the potential to generate employment of 450 man-days per hectare per year (Dhyani *et. al.*, 2013). For example, Poplar is a prominent industrial species which covers less than 2 percent area under agroforestry with 4–5 lakh ha but generates about 90 to 100 million mandays of employment per year. This includes 20 million mandays for felling of trees and their conversion into logs, grading, loading, unloading, stacking and handling timber; and approximately 50 million mandays in wood based industries involved in converting poplar wood into wood based products in addition to 20 million mandays for crop cultivation in India (Dhiman, 2008 and ICFRE, 2016).

According to another estimate of ICAR-CAFRI, the potential of employment generation is 943 million person-days annually from the 25.4 million hectare of existing area under agroforestry (Table-2.1). Presently, the total yearly estimated employment generated under agroforestry system is about 40 percent of the employment generated (2358.4 million person days) under Govt. of India's flagship employment generation scheme, namely 'Mahatma Gandhi National Rural Employment Guarantee Act (MGNREGA)' during 2016–17 at a cost of Rs. 407,822 million incurred on wages under the scheme. It implies that the area expansion under agroforestry system is a viable and relatively cheaper source of employment generation for unskilled and semiskilled labour in rural and semiurban areas in country.

**Table 2.1:** Employment generation potential through agroforestry in India

| Agroforestry system | Area (million ha) | Additional employment generation (person days/ha/year) | Total employment/year (million person days) |
|---|---|---|---|
| Silviculture | 1.8 | 30 | 53.3 |
| Agri-silviculture (irrigated) | 2.3 | 40 | 91.3 |
| Agri-silviculture (rainfed) | 1.3 | 30 | 38.0 |
| Agri-horticulture (irrigated) | 1.5 | 50 | 76.1 |
| Agri-horticulture (rainfed) | 0.5 | 40 | 20.3 |
| Silvipasture | 5.6 | 30 | 167.4 |
| Tree borne oilseeds (TBOs) | 12.4 | 40 | 497.1 |
| Total | 25.4 | 260 | 943.4 |

*Source:* NRCAF (2007) and Dhyani, *et al.* (2009)

## Expected Cost of Cultivation and Financial Viability

The financial analysis of prominent agroforestry systems prevalent in various agro-climatic zones in India was conducted based on the datafrom 'Report of the Task Force on Greening India for Livelihood Security and Sustainable Development' of Planning Commission. All agroforestry systems are financially viable and provide substantive returns to farmers in respective agro-climatic zones. Benefit cost analysis of the system assesses the resource inputs, costs, and benefits over the lifetime of the proposed or existing agroforestry technology. Benefit-cost ratio of some improved agroforestry systems for various Agro-Climatic Zones of India are given in Table-2.2.

**Table 2.2:** Financial analysis of prominent agroforestry systems in various agro-climatic zones, India.

| S. N. | Agro-climatic Zone/ Region | Situation (Rainfed/ Irrigated / Non arable land) | Agro-forestry system | Per-iod (yrs) | Tree (Common name) | Tree (Scientific name) | Crop/ Grass | Total Cost disco-unted @15% (Rs/ha) | NPV (Rs/ha) | B/C ratio |
|---|---|---|---|---|---|---|---|---|---|---|
| 1 | Western Himalayan | Rainfed | Agrihorticulture | 15 | Peach | *Prunus persica* | Maize, Soyabean | 57,144 | 12,039 | 1.21 |
| 2 | Eastern Himalayan | Irrigated | Agrisilviculture | 8 | Kadam | *Anthocephaluscadamba* | Paddy | 43,776 | 24,348 | 1.56 |
| 3 | Lower Gangetic Plains | Irrigated | Agrisilviculture | 10 | Eucalyptus | *Eucalyptus hybrid* | Paddy, Wheat | 86,237 | 114,532 | 2.33 |
| 4 | Lower Gangetic Plains | Rainfed | Agrisilviculture | 15 | Aust. Babool | *Acacia auriculiformis* | Jute | 27,016 | 4,558 | 1.17 |
| 5 | Middle Gangetic Plains | Rainfed | Agrisilviculture | 20 | Shisam | *Dalbergia sissoo* | Seasamu m | 14,360 | 9,030 | 1.63 |
| 6 | Middle Gangetic Plains | Non- arable | Silvipasture | 10 | Bamboo | *Bambusa vulgaris* | Marvel grass | 24,796 | 18,907 | 1.76 |
| 7 | Upper Genetic Plains | Irrigated | Agrisilviculture | 7 | Poplar | *Populus deltoids* | Paddy, Wheat | 82,292 | 190,241 | 3.31 |

| S. N. | Agro-climatic Zone/ Region | Situation (Rainfed/ Irrigated / Non arable land) | Agro-forestry system | Period (yrs) | Tree (Common name) | Tree (Scientific name) | Crop/ Grass | Total Cost discounted @15% (Rs/ha) | NPV (Rs/ha) | B/C ratio |
|---|---|---|---|---|---|---|---|---|---|---|
| 8 | Trans Gangetic Plains | Rainfed | Agrisilviculture | 20 | Babul | *Acacia nilotica* | Bajra, Guar | 17,263 | 27,788 | 2.61 |
| 9 | Eastern Plateau & Hill | Irrigated | Agrisilviculture | 12 | Gumhar | *Gmelina arborea* | Paddy, Linseed | 69,140 | 50,857 | 1.74 |
| 10 | Eastern Plateau & Hill | Non- arable | Agrisilviculture | 15 | Anjan | *Hardwickiabinata* | Dinanath grass | 60,062 | 68,035 | 2.13 |
| 11 | Central Plateau & Hill | Rainfed | Agrihorticulture | 8 | Aonla | *Emblica officinalis* | Ground-nut, Gram | 86,494 | 7,409 | 1.09 |
| 12 | Western Plateau & Hill | Non- arable | Silvipasture | 16 | Siris | *Albiziaamara* | Krishna Siris | 63,236 | 43,768 | 1.69 |
| 13 | Southern Plateau & Hill | Irrigated | Agrisilviculture | 15 | Kapok | *Ceiba pentendra* | Ground-nut | 109,840 | 46,746 | 1.43 |
| 14 | Southern Plateau & Hill | Rainfed | Agrisilviculture | 10 | Imli | *Tamarindusindica* | Chilli | 28,163 | 29,101 | 2.03 |

| S. N. | Agro-climatic Zone/ Region | Situation (Rainfed/ Irrigated / Non arable land) | Agro-forestry system | Period (yrs) | Tree (Common name) | Tree (Scientific name) | Crop/ Grass | Total Cost discounted @15% (Rs/ha) | NPV (Rs/ha) | B/C ratio |
|---|---|---|---|---|---|---|---|---|---|---|
| 15 | West Coast & Ghat | Irrigated | Agrisilviculture | 7 | Beefwood | *Casurinaequisetifolia* | Paddy | 86,889 | 33,907 | 1.39 |
| 16 | West Coast & Ghat | Irrigated | Agrihorticulture | 8 | Coconut | *Cocos nucifera* | Elephant foot, Ginger | 64,387 | 17,650 | 1.27 |
| 17 | West Coast & Ghat | Rainfed | Agrisilviculture | 20 | Teak | *Tectonagrandis* | Sweet Potato | 209,715 | 210,246 | 2.00 |
| 18 | Gujarat Plains & Hill | Rainfed | Agrisilviculture | 20 | Sissoo | *Dalbergia Sissoo* | Castor | 7,080 | 8,037 | 2.14 |
| 19 | Western Dry | Rainfed | Agrisilviculture | 20 | Khejri | *Prosopis cineraria* | Bajra | 35,665 | 54,409 | 2.53 |
| 20 | Western Dry | Non- arable | Silvipasture | 10 | Ber | *Ziziphus nummularia* | Dhaman grass | 37,131 | 19,810 | 1.53 |
| 21 | The Islands | Rainfed | Agrisilviculture | 6 | Coconut | *Cocos nucifera* | Paddy | 37,452 | 6,325 | 1.17 |

*Source:* Planning Commission (2001) Report of the Task Force on Greening India for Livelihood Security and Sustainable Development, Government of India, New Delhi.

## Economics of different agroforestry system in India

There are a number of studies reported from various parts of the country suggesting that agroforestry is more profitable to farmers than agriculture or forestry alone on a particular piece of land. A comprehensive analysis indicated economic viability with internal rate of return (IRR) ranging from 25–68 and B:C ratio of 1.01–4.17 for 24 agroforestry systems in different agro-climatic regions of the country. For example, a B:C ratio of 3.28 (on discounted rate it was 2.61) was reported in a 13 years study of aonla-based agroforestry system in marginal lands under rainfed conditions, which indicated its profitability over other systems (Ram Newaj and Rai, 2005). Deswal, *et al.* (2014) reported 46 per cent higher income from poplar-based agro-forestry system compared to traditional rice-wheat croprotation. Chahal, *et al.*, 2012, recorded highest net income increase of poplar +sugarcane (Rs.64,355/ha/year), followed by poplar+turmeric (Rs. 59,543/ha/year) and lowest net income was recorded in poplar+rainfed wheat (Rs. 18,719/ha/year). Poplar alone gave a net income of Rs. 20,188/ha/year whereas the traditional rice-wheat crop rotation gave net income of Rs. 22,970/ha/year. Chauhan, *et al.* (2015) recorded greater economic returns from *P. deltoids* in block plantation than boundary plantation and sole cropping of rice-wheat (B:C ratio of 3.30, 1.90 and 1.61 respectively). The benefit cost analysis of various systems is given in Table-2.3.

**Table 2.3:** Benefit cost ratio analysis of different agroforestry system in India

| Agroforestry system | B/C ratio | References |
|---|---|---|
| Eucalyptus (10×1.5 m) + crop | 2.20 | Prasad *et al.*, 2010 |
| Eucalyptus (11×1 m) + crop | 2.17 | |
| Eucalyptus (7×1.5 m )+ crop | 2.17 | |
| Eucalyptus (5×1 m) + crop | 2.17 | |
| Eucalyptus (3×2 m) + crop | 2.07 | |
| Sole Eucalyptus | 2.29 | |
| Sole arable cropping | 1.93 | |
| Aonla+Ber+Karonda+Cluster bean + | 3.14 | Arya *et al.*, 2011 |
| BrinjalAonla+Ber+Karonda+Cluster bean + Fallow | 3.07 | |
| Aonla+Ber+Karonda+Moth bean+ Mustard | 3.48 | |
| Aonla+Ber+Karonda+Moth bean + Fallow | 3.09 | |
| Mangium + Pineapple | 2.34 | Nayak *et al.*, 2014 |
| Mangium + Aloevera | 2.10 | |
| Mangium + Kalmegh | 2.20 | |
| Gamhar + Pineapple | 1.65 | |
| Pure eucalyptus plantation (5 yrs) | 3.30 | Rani *et al.*, 2016 |
| *T. grandis* plantation (10 yrs) | 2.06 | |
| *P. deltoids*+ *T. aestivum* (5 yrs) | 2.02 | |

| Agroforestry system | B/C ratio | References |
|---|---|---|
| *Gmelina arborea*+ *Cajanus cajan* | 2.05 | Berry, N. 2017 |
| *Gmelina arborea*+ *Withania somenifera* | 3.80 | |
| *Gmelina arborea*+ *Rauvolfia serpentina* | 6.42 | |
| Sole *Gmelina arborea* | 2.03 | |
| Sole *Cajanus cajan* | 0.72 | |
| Sole *Withania somenifera* | 0.90 | |
| Sole *Rauvolfia serpentina* | 8.09 | |
| Jack fruit+Chili | 4.27 | Ghode, B. D. and Dongre, R. 2017 |
| Jack fruit+Tomato | 3.41 | |
| Sole Chili | 2.85 | |
| Sole Tomato | 1.78 | |
| FS+Lac+Cc-Le | 7.08 | Berry, N. and Singh, N. 2017 |
| FS+Lac | 5.16 | |
| FM+Lac+CC-Le | 4.68 | |
| FM+Lac | 3.03 | |
| LE | 2.96 | |
| CC | 2.15 | |
| *Where*, FS=*F.semialata*, FM=*F.macrophylla*, CC=*Cajanus cajan* and LE= *Lycopersiconesculantom* | | |
| *B. vulgaris*+Sudan grass | 4.40 | Puran *et al.*, 2018. |
| *B. vulgaris*+Napier grass | 2.60 | |
| *B. vulgaris*+Guinea grass | 3.74 | |
| Sole *B. vulgaris* | 2.67 | |
| Sole Sudan grass | 3.70 | |
| Sole Napier grass | 0.72 | |
| Sole Guinea grass | 2.11 | |
| Mango+Mustard | 3.61 | Sahu *et al.*, 2019 |
| Mango+Linseed | 3.12 | |
| Sole Mango | 1.86 | |
| Sole Mustard | 1.76 | |
| Sole Linseed | 1.28 | |
| *M. dubia* +Pearl millet (control) (6 year old) | 5.36 | Chandana *et al.*, 2020 |
| *M. dubia* + Pearl millet 100% RDF through normal urea | 6.20 | |
| *M. dubia*+Pearl millet 100% RDF through neem-coated urea | 7.24 | |
| *M. dubia*+Pearl millet RD N + 25% N through poultry manure | 5.48 | |
| *M. dubia* + Pearl millet 75% RD N + 25% N through FYM | 5.97 | |
| *M. dubia* + Pearl millet 75% RD N + PGLM @ 10 t /ha | 5.79 | |
| *M. dubia* + Pearl millet 75% RD N + Azotobacter@ 500 g/ ha | 5.39 | |
| Sole Pearl millet without trees (80–40–30 NPK kg/ ha) | 2.96 | |

There are a number of studies indicating profitability of the agroforestry systems. However, in most of the economic analysis of agroforestry systems, only the cost of inputs and outputs in material terms have been taken into account. If the cost of goods and services produced by the agroforestry systems in temporal and spatial dimensions has also been considered for the overall economic analyses then the benefits of agroforestry system would have become much higher.

**Potential of Agro-forestry sector ALUS (Alternate Land-Use Systems) esp. in barren/waste lands & farm bunds**

1. *Acacia fortilis* plantation (3m X 3m spacing) is most appropriate for getting maximum biomass and monetary return up to Rs. 75,000 to 1.25 lakh / ha (in 2016–17 prices).
2. Minor fruit crops and medicinal & aromatic plants as crop diversification or cultivation in marginal lands and field bunds can improve income levels up to 40 per cent.
3. Tree species of 4–6 years of rotation age such as Ailanthus (match stick tree), *Melia dubia* have the potential to generate 40% additional monetary return.
4. Silvi-pasture system (Amla + *Dichanthium annulatum*) can improve fodder supply @ 20%, increment Amla yields by 20% and enhance net returns up to Rs. 6500 / ha/ annum.
5. *Ailathus excelsa* (6.0 × 3.0 mt spacing) based agri-silviculture system (Ardu + Green gram) can increment farmers' income by 25%.
6. India's paper demand to increase by 53% by 2022: Trees with good pulpwood materials are heavily demanded by paper and pulp industry.
7. Advantage for Gujarat: Semi-arid districts (Anand, Nadiad, Mehsana and Gandhinagar) are with high tree cover and are already wood surplus districts.
8. Silvi-medicinal systems have the next big income potential: Teak + *Ocimum spp* such as *O. tenuiflorum, O. gratissimum and O. basilicum* increase net returns by Rs. 80,000/ha (2014–15 prices). Farmers can also get additional income from such land-uses till teak trees come to harvest.
9. *Melia dubia* trees on farm bunds: 150–170 trees / hectare: 13–14 years rotation can yield timber for door & furniture of Rs 15,000 per tree value and 2–3 year rotation can yield 10 to 12.5 tons of pulpwood of Rs 4,000 per ton value.

## Conclusion

Income of farmers could be enhanced substantially by allied activities such as agroforestry, horticultural, medicinal and aromatic crops cultivation, dairy and fisheries, value addition to harvested biomass of plant and livestock origin in the production catchment and using crop residues and processing byproducts for feed and manures to enhance livestock and soil-health and thereby better productivity with good quality produce, which in turn will fetch more income to farmers and improve the health of the consumers. The Government of India has taken a number of initiatives to increase farm income, stabilize production and consequently, improve small farm productivity. Agroforesstry system approach involving synergic blending of trees, crops, horticulture, fodder, fisheries, etc. seems viable option to provide regular income and at site employment to small-land holders.

## References

Arya R, Awasthi OP, Singh J and Singh B 2011 Cost benefit analysis under fruit based multiple cropping systems. *Progressive Horticulture* **43**(1) 72–75.

Berry N. 2017 Economic analysis of *Gmelina* based silvi-agri-medicinal system. Agroforestry Division, Tropical Forest Research Institute, Mandla Road, Jabalpur (M.P.). Paper presented in annual group meeting of AICRP- Agroforestry, 28–30 July 2018, Birsa Agricultural University, Ranchi, Jharkhand.

Berry, N. and Singh, N. 2017 Economic analysis of *F. semialata* based silvi-agri-lac system during second year. Tropical Forest Research Institute, Jabalpur, (M.P.). Paper presented in annual group meeting of AICRP- Agroforestry, 28–30 July 2018, Birsa Agricultural University, Ranchi, Jharkhand.

Chahal D, Ahmad A, and Bhatia JN 2012 Assessment of agroforestry based two-tier cropping system in Ambala district of Haryana. *Agriculture Update* 7(3/4) 210–213.

Chandana P, Lata AM, Khan MAA and Krishna A 2020 Climate change smart option and doubling farmer's income through *Melia dubia* based agri-silviculture system. *Current Science* **118**(3) 444–447.

Chauhan SK, Sharma R, Singh B and Sharma SC 2015 Biomass production, carbonsequestration and economics of on-farm poplar plantations in Punjab. *Indian Journal of Applied and Natural Science* **7**(1) 452–458.

Deswal AK, Pawar N, Raman RS, Yadav JN and Yadav VPS 2014 Poplar (*Populus deltoides*) based agroforestry system: a case study of Southern Haryana. *Annals of Biology* **30**(4) 699–700.

Dhyani SK, Handa AK and Uma 2013 Area under agroforestry inIndia: An assessment for present status and futureperspective. *Indian J. of Agroforestry* **15**(1) 1 11.

Doubling Farmers' Income, 2018 Farm linked Activities and Secondary Agriculture. Department of Agriculture, Cooperation and Farmers' Welfare, Ministry of Agriculture & Farmers' Welfare. Volume IX, 75–82 pages.

FSI 2013 India State of Forest Report. Ministry of Environment and Forests, Dehradun, India.

Ghode BD and Dongre R 2017 Economics of Jack fruit based Agrisilviculture system, Department of Forestry, College of Agriculture, JNKVV, Jabalpur (M. P.). Paper presented in annual group meeting of AICRP- Agroforestry, 28–30 July 2018, Birsa Agricultural University, Ranchi, Jharkhand.

Nayak MR, Behera LK, Mishra PJ and Bhola N 2014 Economics and yield performance of some short duration fruit and medicinal crops under agrisilvicultural system in rainfed uplands of Odisha. *Journal of Applied and Natural Science* **6**(1) 274–278.

Planning Commission 2001 Report of the Task Force on greening India for livelihood security and sustainable development. Planning Commission, Government of India, 254 page.

Prasad JVNS, Korwar GR, Rao KV, Mandal UK, Rao CAR, Rao GR, Ramakrishna YS, Venkateswarlu B, Rao SN, Kulkarni HD and Rao MR 2010 Tree row spacing affected agronomic and economic performance of Eucalyptus-based agroforestry in Andhra Pradesh, Southern India. *Agroforest System* **78** 253–267. DOI 10.1007/s10457–009–9275–1

Puran A, Malik MS, Oroan PR and Kumar A 2018 Study on Growth and Economics of Bamboo Based Silvipastoral System. *Int. J. Curr. Microbiol. App. Sci.*, **SI7** 4273–4277

Ram Newaj and Rai P 2005 Aonla-based agroforestry system: a source of higher income under rainfed conditions. *Indian Farm.* **55** 24–27.

Rani S, Rajasekharan A, Benbi DK. and Chauhan SK 2016 Economic Evaluation of Different Land Use Systems in North Western Region of Punjab, India. *Forest Res.*, **6**(1) 192–199.

Report of the Committee on Doubling Farmers' Income 2018 Farm linked Activities and Secondary Agriculture. Volume IX.

CHAPTER - 3

# Agroforestry Policy and Schemes

## NATIONAL AGROFORESTRY POLICY-2014

The World Congress on Agroforestry was organized in February 2014 at New Delhi with the theme 'Trees for Life' to have an outlook to any constraint that might restrict the adoption of agroforestry practices. More than a thousand delegates from around 80 countries attended the Congress, along with the President of India launched the much-needed National Agroforestry Policy (NAP), 2014 first of its kind in the world. NAP, 2014 is a recognition to the professionals who have spent half a century of research and documentation and thereby made agroforestry an integrated science. Moreover, NAP, 2014 is a path-breaker in making agroforestry an instrument for transforming the lives of the rural farming population, protecting ecosystem and ensuring food security through sustainable means.

The major highlights of the Policy are: establishment of institutional setup at the national level to promote agroforestry under the mandate of the Ministry of Agriculture, Government of India; simplify regulations related to harvesting, felling and transportation of trees grown on farmlands; ensuring security of land tenure and creating a sound base of land records and data for developing an market information system (MIS) for agroforestry; investing in research, extension and capacity building and related services; access to quality planting material; institutional credit and insurance cover to agroforestry practitioners; increased participation of industries dealing with agroforestry produce, and strengthening marketing information system for tree products. Initially 20 important multipurpose tree species have been identified at the national level to be exempted from all restrictions related to harvesting, transportation and marketing grown under agroforestry systems. The Policy also suggests massive extension programmes in order to broadcast the outcomes of intensive R & D activities in the field of agroforestry.

### Objectives

- Encourage and expand tree plantation in complementarity and integrated manner with crops and livestock to improve productivity, employment, income and livelihoods of rural households, especially the small holder farmers.

- Protect and stabilize ecosystems, and promote resilient cropping and farming systems to minimize the risk during extreme climatic events.
- Meet the raw material requirements of wood based industries and reduce import of wood and wood products to save foreign exchange.
- Supplement the availability of agroforestry products, such as the fuel-wood, fodder, non-timber forest produce and small timber of the rural and tribal populations, thereby reducing the pressure on existing forests.
- Complement achieving the target of increasing forest/tree cover to promote ecological stability, especially in the vulnerable regions.
- Develop capacity and strengthen research in agroforestry and create a massive people's movement for achieving these objectives and to minimize pressure on existing forests.

## Strategy

1. Establishment of Institutional Setup at National level to promote Agroforestry
2. Simple regulatory mechanism
3. Development of a sound database & information system
4. Investing in research, extension and capacity building and related services
5. Improving famers' access to quality planting material
6. Providing institutional credit and insurance cover for agroforestry
7. Facilitating increased participation of industries dealing with agroforestry produce
8. Strengthening farmer access to markets for tree products
9. Incentives to farmers for adopting agroforestry
10. Promoting sustainable agroforestry for renewable biomass based energy

## Sub-Mission on Agroforestry-2016

"***Har Medh Par Pedh***"-Tree Planting with Agricultural Crops

The Government of India has implemented the Sub-Mission on Agroforestry (SMAF) in 2016 under the National Mission for Sustainable Agriculture (NMSA) to promote tree plantation on farmland along with the productivity enhancement of crops and cropping system. The scheme had been implemented in the states having liberalised transit regulations for transport of timber. This would not only help in reducing the effects of climate change, increase soil biodiversity, but it would also provide a source of income for farmers.

The principal objective of implementing this scheme is to address the development of agroforestry practices for the increased crop production and economic gain in the country.

## Objectives

- Tree plantation on farm land "Har Medh Par Ped", along with crops/ cropping system.
- Multipurpose tree species with short, medium and long term returns is planted, so that farmers may get additional income at regular intervals. These can include fruits, fodder, medicinal, timber & aromatic tree species.
- The scheme is being implemented in the States which exempted selected tree species from transit permit.
- To ensure the production of quality planting material such as seeds, seedlings, hybrids, clones, improved varieties, shoot tips, root tips and pollen grains.
- To popularise various agroforestry methods and practices which are suitable for different agro-ecological zones and land use conditions.
- To provide extension and building capacity support for the agroforestry research sector.
- To build opportunities for small scale forest based enterprises and also creates a database on the area about the status of soil organic carbon, information and knowledge about the agroforestry.

### Mission Interventions/Activities under SMAF

1. Nursery Development for the quality planting material(NDQPM)
2. Peripheral and Boundary Plantation (PBP)
3. Low Density Plantation on the Farm Lands (LDPFL)
4. High Density Block Plantation (HDBP)
5. Demonstration of the Agroforestry Models
6. Capacity Building and Trainings.

Under the agroforestry scheme (financed jointly by the Central Government and the concerned State Governments in 60:40 ratio), farmers are eligible to get a subsidy of Rs. 35 per tree grew (Rs. 35 to be contributed by the farmer). The Additional Secretary sought the help of Paper Mills in spreading awareness amongst farmers engaged in agroforestry and helps them get registered under the scheme.

## Compensatory Afforestation Fund Management and Planning Authority (CAMPA)-2016

Compensatory Afforestation Fund Act (CAMPA), 2016 came into force from 30 September 2018. The Act established a National Compensatory Afforestation Fund under the Public Account of India and State Compensatory Afforestation Fund under the Public Account of each state. The payments made for compensatory afforestation, net present value and others related to the project will be deposited in the fund. The State Funds will receive 90% of the payments while National Fund will receive remaining 10%. These funds will be regulated by State and National CAMPA. MoEFCC handed over Rs. 47,436 crore to various states to take up afforestation and conservation activities as per the provisions of CAF Act and Rules. The Ministry also stressed that the fund had to be used for important needs such as Compensatory Afforestation, Catchment Area Treatment, Wildlife Management, Assisted Natural Regeneration, Forest Fire Prevention and Control Operations, Soil and Moisture Conservation Works in the forest, Improvement of Wildlife Habitat, Management of Biological Diversity and Biological Resources, Research in Forestry and Monitoring of CAMPA works and others.

Many development and industrial projects such as erection of dams, mining, and construction of industries or roads require diversion of forest land. Any project proponent, government or private must apply for forest clearance from Ministry of Environment Forests and Climate Change (MoEFCC), before the conversion of land take place. This proposal is to be submitted through the concerned forest department of the state government. If clearance is given, then compensation for the lost forest land is also to be decided by the ministry and the regulators.

The Hon'ble Supreme Court on 10th July 2009 issued orders that there will be a Compensatory Afforestation Fund Management and Planning Authority (CAMPA) as National Advisory Council under the chairmanship of the Union Minister of Environment Forests & Climate Change for monitoring, technical assistance and evaluation of compensatory afforestation activities.

## Objectives of CAMPA

Compensatory Afforestation Fund Management and Planning Authority (CAMPA) are meant to promote afforestation and regeneration activities as a way of compensating for forest land diverted to non-forest uses. National CAMPA Advisory Council has been established as per orders of The Hon'ble Supreme Court with the following mandate:

- Lay down broad guidelines for State CAMPA.
- Facilitate scientific, technological and other assistance that may be required by State CAMPA.
- Make recommendations to State CAMPA based on a review of their plans and programmes.
- Provide a mechanism to State CAMPA to resolve issues of an inter-state or Centre-State character.

## Green Watch-2017

**Green Watch: An eye on the green** A web portal has been put in place to monitor plantation and the growth of seedlings. Till May 2017, 28 states uploaded 84,326 works under CAMPA, which has been analysed.

## Intended Nationally Determined Contributions-2015

Globally committed to create a new international climate agreement by the conclusion of the United Nation Framework Convention on Climate Change (UNFCCC), Conference of the Parties (COP-21) in Paris in December 2015. In preparation, countries have agreed to publicly outline what post-2020 climate actions they intend to take under a new international agreement, known as their Intended Nationally Determined Contributions (INDCs). The INDCs will largely determine whether the world achieves an ambitious 2015 agreement and is put on a path toward a low-carbon, climate-resilient future.

India has submitted its Intended Nationally Determined Contribution (INDC) to the United Nations Framework Convention on Climate Change.

**Salient features of India's INDC**

- To put forward and further propagate a healthy and sustainable way of living based on traditions and values of conservation and moderation.
- To adopt a climate-friendly and a cleaner path than the one followed hitherto by others at corresponding level of economic development.
- To reduce the emissions intensity of its GDP by 33 to 35 per cent by 2030 from 2005 level.
- To achieve about 40 per cent cumulative electric power installed capacity from non-fossil fuel based energy resources by 2030, with the help of transfer of technology and low cost international finance, including from Green Climate Fund.
- To create an additional carbon sink of 2.5 to 3 billion tons of $CO_2$ equivalent through additional forest and tree cover by 2030.
- To better adapt to climate change by enhancing investments in development programmes in sectors vulnerable to climate change, particularly agriculture, water resources, Himalayan region, coastal regions, health and disaster management.
- To mobilize domestic and new and additional funds from developed countries to implement the above mitigation and adaptation actions in view of the resource required and the resource gap.
- To build capacities, create domestic framework and international architecture for quick diffusion of cutting edge climate technology in India and for joint collaborative R&D for such future technologies.

**Key elements and focus areas of India's INDC**

1. Sustainable Lifestyles
2. Cleaner Economic Development
3. Reducing Emission intensity of Gross Domestic Product (GDP)
4. Increasing the Share of Non Fossil Fuel Based Electricity
5. Adaptation
6. Mobilizing Finance
7. Technology Transfer and Capacity Building
8. Enhancing Forests Carbon Sink

## Goals

- To Create additional carbon sink of 2.5–3.0 billion tons of $CO_2$ equivalent through additional forest and tree cover (increase of about 680–817 million tons of carbon stock).

I. **Emphasis on India's Plans to enhance its carbon sink:**

- Full implementation of Green India Mission
- India has recently formulated Green Highways (Plantation & Maintenance) Policy to develop 140,000 km long "tree-line" with plantation along both sides of national highways. 1% of total civil cost of projects is to be set aside to implement the policy.
- Plantation along Rivers: part of the Namami Gange Mission.

II. **Finance Commission (FC) Incentive for creation of carbon sink: devolution of funds to states from federal pool (attaches 7.5% weight to area under forest).**

III. **Reduction in consumption of wood/ biomass as fuel.**

IV. **Funds from Compensatory Afforestation Fund Management and Planning Authority (CAMPA): USD 6 billion proposed to be given to States.**

V. **Other Policies including**

- REDD+
- National Agro-forestry Policy (NAP)
- Joint Forest Management
- National Afforestation Programme

## National Adaptation Fund for Climate Change-2015

The National Adaptation Fund for Climate Change (NAFCC) was established in August, 2015 to meet the cost of adaptation to climate change for the State and Union Territories of India that are particularly vulnerable to the adverse effects of climate change. Government has set up a budget provision of Rs.350 crores for the year 2015–16 and 2016–17, with an estimated requirement of Rs. 181.5 crores for financial year 2017–18 for NAFCC. The project under NAFCC prioritizes the needs that builds climate resilience in the areas identified under the SAPCC (State Action Plan on Climate Change) and the relevant Missions under NAPCC (National Action Plan on Climate Change).

NABARD has been designated as National Implementing Entity (NIE) for implementation of adaptation projects under NAFCC by Govt. of India. Under this arrangement, NABARD would perform roles in facilitating identification of project ideas/concepts from State Action Plan for Climate Change (SAPCC), project formulation, appraisal, sanction, disbursement of fund, monitoring & evaluation and capacity building of stakeholders including State Governments.

The National Adaptation Fund for Climate Change (NAFCC) is a Central Sector Scheme which was set up in the year 2015–16. The overall aim of NAFCC is to support concrete adaptation activities which mitigate the adverse effects of climate change. The activities under this scheme are implemented in a project mode. The projects related to adaptation in sectors such as agriculture, animal husbandry, water, forestry, tourism etc. are eligible for funding under NAFCC.

## Objective

1. Funding concrete adaptation projects/programmes aligned with the relevant Missions under National Action Plan on Climate Change (NAPCC) and the State Action on Climate Change (SAPCCs) in agriculture, horticulture, agroforestry, environment, allied activities, water, forestry, urban, coastal and low-lying system, disaster management, human health, marine system, tourism, habitat sector and other rural livelihood sectors to address climate change related issues.
2. Preparing and updating climate scenario, assessing vulnerability and climate impact assessment.
3. Capacity building of various stakeholders on climate change adaptation and project cycle management and developing knowledge network.
4. Mainstreaming the approaches/learnings from project/programme implementation through knowledge Management.

## National Mission for Sustainable Agriculture (NMSA)-2014

NMSA is envisaged as one of the eight Missions outlined under National Action Plan on Climate Change (NAPCC), NMSA aim at promoting Sustainable Agriculture through climate change adaptation measures. The major thrust is enhancing agriculture productivity especially in rainfed areas focusing on integrated farming, soil health management, and synergizing resource conservation. Besides, NMSA also a committed target to fulfill National and International commitments on Sustainable Development Goals (SDG) & Intended Nationally Determined Contribution (INDC). All the components of NMSA such as Rainfed Agriculture, Soil Health Management, Organic Farming, etc. have significant role in achieving SDGs & INDC.

NMSA as a programmatic intervention caters to Mission Deliverables that focuses mainly on conservation agriculture to make farm sector more productive, sustainable, remunerative and climate resilient by promoting location specific integrated/composite farming systems; soil and moisture conservation measures; comprehensive soil health management and mainstreaming rainfed technologies.

## NMSA has the following schemes

1. Rainfed Area Development (RAD): RAD is being implemented by RFS Division.
2. Sub Mission on Agro Forestry (SMAF): SMAF is being implemented by NRM Division.

3. Paramparagat Krishi Vikas Yojana (PKVY): PKVY is being implemented by INM Division.
4. Soil and Land Use Survey of India (SLUSI): Being implemented by INM Division.
5. National Rainfed Area Authority (NRAA): Being implemented by RFS Division.
6. Mission Organic Value Chain Development in North Eastern Region (MOVCDNER): Being implemented by INM Division.
7. National Centre of Organic Farming (NCOF): Being implemented by INM Division.
8. Central Fertilizer Quality Control and Training Institute (CFQC&TI): Being implemented by INM Division.

Rainfed Area Development (RAD):-RAD focuses on Integrated Farming System (IFS) for enhancing productivity and minimizing risks associated with climatic variability's. Under this system, crops/cropping system is integrated with activities like horticulture, livestock, fishery, agroforestry, apiculture, forestry etc. to enable farmers not only in maximizing farm returns for sustaining livelihood, but also to mitigate the impacts of drought, flood or other extreme weather events with the income opportunity from allied activities during crop damage. For the Year 2019–20, budget provision of Rs. 250.0 crore has been made for implementation of the programme.

## National Action Plan on Climate Change-2008

The National Action Plan on Climate change (NAPCC) was formally launched on 30 June, 2008. The NAPCC identifies measures that promote development objectives while also yielding co-benefits for addressing climate change effectively.

There are eight "National Missions" which form the core of the National action plan. They focus on promoting understanding of climate change, adaptation and mitigation, energy efficiency and natural resource conservation.

The eight missions are:

1. National Solar Mission.
2. National Mission for Enhanced Energy Efficiency.
3. National Mission on Sustainable Habitat.
4. National Water Mission.
5. National Mission for Sustaining the Himalayan Ecosystem.
6. National Mission for a Green India- The NAPCC also aims at afforestation of 6 million hectares of degraded forest lands and expanding forest cover from 23 to 33% of India's territory. Recently initiatives like Green India Mission (GIM) aim to further increase the forest/tree cover to the extent of 5 million hectares (m ha) and improve quality of forest/tree cover on another 5 m ha of

forest/non-forest lands along with providing livelihood support. It is expected to enhance carbon sequestration by about 100 million tons CO2 equivalent annually.

7. National Mission for Sustainable Agriculture.
8. National Mission on Strategic Knowledge for Climate Change.

## National Bamboo Mission-2006

In October 2006, the Government of India (GOI) had launched the National Bamboo Mission (NBM) on the basis of the National Mission on Bamboo Technology and Trade Development Report, 2003. The NBM's key objective was to address issues relating to the development of the bamboo industry in the country, provide a new impetus and direction of India's considerable potential in bamboo production.

Bamboo is a versatile crop. It can be used in 1,500 different ways including as food, a substitute for wood, building and construction material, for handicrafts and paper. Around 80 per cent of bamboo forests lie in Asia with India, China and Myanmar. India is the world's second largest cultivator of bamboo after China, with 136 species and 23 genera spread over 16 million hectares (ISFR-2019). According to the Union Ministry of Agriculture and Farmer Welfare, India's annual bamboo production is estimated at 3.23 million tons. However, despite all this, the country's share in the global bamboo trade and commerce is only 4 per cent.

The NBM's totals of 1,466 nurseries have been established to supply quality planting material. In different states, 61,126 farmers and 12,710 field functionaries have been trained in the area of nursery management and bamboo plantation. In addition, 39 bamboo wholesale and retail markets near villages, 29 retail outlets and 40 bamboo bazaars have been established. The Mission is expected to establish about 4000 treatment/ product development units and bring more than 100000 ha area under plantation during the period 2018–19 and 2019–20.

In April 2018, a restructured NBM was approved by the Government of India with an investment of Rs 1290 crore in the coming 2 years. The restructured NBM will aim to support the development of the entire value chain of the bamboo sector starting from planting material, plantation, creation of facilities for collection, aggregation, processing marketing, micro, small & medium enterprises, skill development and brand building initiative in a cluster approach mode. According to the State of India's Environment 2017, nearly 30 per cent of India's land is degraded. With its unique ability to stitch and repair damaged soils, bamboo is ideal for rehabilitating degraded soil. The short fall in bamboo plantation would have adverse impact in reclaiming degraded land as well as the Greening India programme.

This will contribute to doubling of farmers' income and also generate more employment opportunities for skilled and unskilled workers, especially youths in rural areas. To facilitate the benefit flow to the farmers, bamboo outside forest areas has been excluded from the definition of tree by amending Section 2(7) of the Indian Forest Act, 1927 by the GOI in November 2017.

### Objective

1. To promote the growth of the bamboo sector through as an area based regionally differentiated strategy
2. To increase the coverage of area under bamboo in potential areas, with improved varieties to enhance yields
3. To promote marketing of bamboo and bamboo based handicrafts
4. To establish convergence and synergy among stake-holders for the development of bamboo
5. To promote, develop and disseminate technologies through a seamless blend of traditional wisdom and modern scientific knowledge
6. To generate employment opportunities for skilled and unskilled persons, especially unemployed youths.

### Strategy

- Adopt a coordinated approach covering production and marketing to assure appropriate returns to growers/producers.
- Promote Research and Development (R&D) of varieties and technologies for enhanced production.
- Enhance acreage (in forest and non-forest areas) and productivity of bamboo through varietal change and improved agriculture practice.
- Promote partnership, convergence and synergy among R&D and marketing agencies in public as well as private sectors, at all levels.
- Promote where appropriate, cooperatives and self-help groups to ensure support and adequate returns to farmers.
- To generate employment opportunities for skilled and unskilled persons, especially unemployed youths.
- Set up National, State and sub-State level structures, to ensure adequate returns for the produce of the farmers and eliminate middlemen, to the extent possible.

### National Horticulture Mission-2005

A National Horticulture Mission has been launched in 2005–06 as a Centrally Sponsored Scheme to promote holistic growth of the horticulture sector through an area based regionally differentiated strategies. The scheme is fully funded by the Government and different components proposed for implementation financially supported on the scales laid down. Presently, India is the 2nd largest producer of fruits & vegetables in the world.

National Horticulture Mission (NHM) is being implemented in all the States and Union Territories of India except the North Eastern States, Himachal Pradesh,

Jammu & Kashmir and Uttaranchal (for which a separate Technology Mission for integrated development of horticulture exists) to promote holistic growth of the horticulture sector covering fruits, vegetables, roots and tuber crops, mushroom, spices, flowers, aromatic plants, cashew and cocoa. Programme for the development of coconut will be implemented by the Coconut Development Board (CDB), independent of the Mission. This will be a centrally sponsored scheme in which Government of India shall provide 100% assistance to the State Mission during Tenth plan. During the XI plan, the Government of India assistance will be 85% with 15% contribution by the State Governments.

## Objective

1. Enhance development of horticulture in our country and expanding related activities to the fullest so that maximum tapping of the horticultural resources is possible through the advanced growth of all such products namely crops, vegetables, fruits, spices, flowers and medicinal and aromatic plantations
2. Implementation of an area specific approach for the comprehensive growth of the horticulture sector
3. To increase the horticulture productions in general which in turn will reflect on the enhanced financial security and nutritional aspects of population engaged in different sectors of horticulture
4. To enhance coordination and collaboration amongst the ongoing multiple programs with the common goal of development of the horticultural sector
5. To enhance production of the horticultural sector through a successful combination of time tested practices and modern technologies in this field
6. Horticultural sector is an employment intensive sector. The plans outlined in the NHM aimed at providing employment opportunities especially for the unemployed youths, not to mention inclusion of skilled and unskilled labour
7. The horticulture sector is also an export oriented sector. The overall development of this sector will also ensure an increase in exports from this sector

## National Agriculture Policy-2000

On 28 July 2000, Government of India made public a National Agriculture Policy (NAP) envisaging over 4 per cent annual growth through efficient use of resources and technology and increased private investment while emphasizing on price protection to farmers in the WTO regime.

The broad objectives of NAP is to actualise the vast untapped growth potential of Indian agriculture, strengthen rural infrastructure to support faster agricultural development, promote value addition, accelerate the growth of agro business, create employment in rural areas, secure a fair standard of living for the agricultural population, discourage migration to urbanareas and face the challenges arising out of economic liberalisation and globalisation.

NAP aims to attain a growth rate in excess of four percent per annum in the agriculture sector that is based on efficient use of resources and conserves our soil, waterand bio-diversity and growth that is demand driven and caters to domestic markets and maximises benefits from exports of agricultural products in the face of the challenges arising from economic liberalisation and globalisation.

**National Agricultural Policy (2000) has taken the following important objectives:**

1. **Sustainability in Agriculture**

   The new policy seeks to introduce economically viable, technically sound, environmentally non-degrading and non-hazardous and socially acceptable use of natural resources of the country for promoting the concept of sustainable agriculture.

   - To use unutilized barren wastelands for agriculture and afforestation.
   - To contain biotic pressures on land and to control indiscriminate division of agricultural lands for non-agricultural uses.
   - To enhance cropping intensity through multi-cropping and intercropping.
   - To emphasize rational use of ground and surface water so that over-exploitation of ground water resources can be checked. To adopt better technologies such as drip and sprinkler irrigation system so as to arrange more economic and efficient use of water.
   - To adopt vigorously a long-term perspective plan for sustainable rain-fed agriculture by adopting watershed approach and water harvesting method for development of two-thirds of cropped area of the country which is dependent on rainfall.
   - Involvement of farmers and landless labourers will be sought in the development of pastures/ forestry programmes on huge public wasteland by providing adequate financial incentives and entitlement of trees and pastures.
2. Food and Nutritional Security.
3. Generation and Transfer of Technology.
4. Inputs Management.
5. Incentives for Agriculture.
6. Investments in Agriculture.
7. Institutional Structure.
8. Risk Management.
9. Management Reforms.
10. Employment in agriculture sectors.

## National Medicinal Plants Board-2000

To promote medicinal plants sector, the Government of India has set up National Medicinal Plants Board (NMPB) on 24 November 2000. Currently the board is located in Ministry of AYUSH (Ayurveda, Yoga & Naturopathy, Unani, Siddha & Homoeopathy), Government of India, has laid emphasis on integrating medicinal plants and trees along with agricultural crops. Establishment of silvi-pastoral systems and fodder blocks/ banks has been envisaged in various schemes operated by the Animal Husbandry Department.

## National Afforestation and Eco-development Board-1992

The National Afforestation and Eco-Development Board (NAEB), established in August 1992 under the Ministry of Environment, Forests and Climate Change, Government of India, is responsible for promoting afforestation, tree planting, ecological restoration and eco-development activities in the country, with special attention to the degraded forest areas and lands adjoining the forest areas, national parks, sanctuaries and other protected areas as well as the ecologically fragile areas like the Western Himalayas, Aravallis, Western Ghats, etc. and also promoting agroforestry practices on farm and wastelands.

The detailed role and functions of the NAEB are given below.

- Evolve mechanisms for ecological restoration of degraded forest areas and adjoining lands through systematic planning and implementation, in a cost effective manner.
- Restore through natural regeneration or appropriate intervention the forest cover in the country for ecological security and to meet the fuel wood, fodder and other needs of the rural communities.
- Restore fuel wood, fodder, timber and other forest produce on the degraded forest and adjoining lands in order to meet the demands for these items.
- Sponsor research and extension of research findings to disseminate new and proper technologies for the regeneration and development of degraded forest areas and adjoining lands.
- Create general awareness and help foster people's movement for promoting afforestation and eco-development with the assistance of voluntary agencies, non-government organisations, Panchayati Raj institutions and others and promote participatory and sustainable management of degraded forest areas and adjoining lands.
- Coordinate and monitor the Action Plans for afforestation, tree planting, ecological restoration and eco-development.
- Undertake all other measures necessary for promoting afforestation, tree planting, ecological restoration and eco-development activities in the country.

## National Bank for Agriculture and Rural Development (NABARD)-1982

National Bank for Agriculture and Rural Development (NABARD) came into existence on 12 July 1982 by transferring the agricultural credit functions of RBI and refinance functions of the Agricultural Refinance and Development Corporation (ARDC). It was dedicated to the service of the nation by the late Prime Minister Smt. Indira Gandhi on 05 November 1982. Set up with an initial capital of Rs.100 crore, its' paid up capital stood at Rs. 12,580 crore as on 31 March 2019. Consequent to the revision in the composition of share capital between Government of India and RBI, NABARD today is fully owned by Government of India.

### Mission

Promote sustainable and equitable agriculture and rural development through participative financial and non-financial interventions, innovations, technology and institutional development for securing prosperity.

## Bharatiya Agro Industries Foundation (BAIF)-1967

The BAIF Development Research Foundation is an award-winning charitable organisation based in Urali Kanchan near Pune in Maharashtra, India, that pioneers agricultural development. It was founded in 1967 by Manibhai Desai as the Bharatiya Agro Industries Foundation.

Wadi programme initiated by BAIF Development Research Foundation in south Gujarat in the 1980s, further expanded to different tribal regions of India. The programme which ensures food security through agri-horti-forestry and better quality of life through women improvement, community health, literacy and development of grass root level people's organizations is being expanded throughout the country with the support from the Ministry of Tribal Affairs and NABARD.

BAIF has assisted over 1.81 lakh families to establish 68586 ha wadi, spread over Maharashtra, Gujarat, Karnataka, Uttar Pradesh, Uttarakhand, Rajasthan, Madhya Pradesh, Chhattisgarh, Bihar, Andhra Pradesh and Jharkhand states. For example, horticulture crops such as amla, mango ($10 \times 10$ m$^2$ spacing), and cashew ($7 \times 7$ m$^2$ spacing) with intercrops grown in these spaces and trees like gliricidia, subabul planted in closer spacing in the border in an area of 0.4–10 ha. Based on the initial trials, the Tribal Resource Centre sponsored by NABARD has taken up new initiatives to introduce new cashew varieties such as NRCC-2 and Vengurla-1, Amrapali variety of mango for production and processing, promotion of pineapple, papaya and perennial castor, drought tolerant chick pea and sun hemp as alternative intercrops in wadi in semi-arid regions. 400 wadi families in Chandrapur district of Maharashtra and Betul district of Madhya Pradesh, have taken up lac cultivation on the trees of Palas, Ber and Kusum grown on the farm bunds and common lands. Over 3000 Mahua grafts have been distributed to establish on barren lands to reduce the gestation period. Series of field manuals and documentary films have been produced for training and empowerment of field functionaries and farmers engaged in agri-horti-forestry.

## Agroforestry Plantation Insurance

There is a need to expand the tree insurance initiatives taken in Tamil Nadu for 'Agroforestry Plantation Insurance' by United India Insurance, Chennai, for ensuring financial protection against natural calamities (flood and cyclone), wild animal damage, and pest and disease for the species that are widely cultivated in the state, including casuarina, eucalyptus, sissoo, *G. arborea, M. dubia, A. excelsa* and subabul at the national level covering all major tree species grown under agroforestry.

## Agroforestry Insurance Scheme

In Kerala, insurance scheme for coconut and rubber is in existence. For mature and immature plantations of rubber, provision for compensation on account of natural calamities will be provided with replacement cost plus future returns of present value arising out of loss or death of the plant. Agricultural Insurance Company India Limited is providing biofuel tree/plant insurance for the species such as *Jatropha curcas, Pongamia pinnata, Azadirachta indica, Bassia latifolia, Callophyllum inophyllum* and *Simarouba glauca* at the premium cost of 125–150% of the input cost. Unlike crop insurance, the credit cover mentioned in the Policy is positively expected to receive widespread attention and support from the farmers.

## Agroforestry Contract Farming Model

Contract farming model in Tamil Nadu: a value chain Tamil Nadu Agricultural University (TNAU), Coimbatore has demonstrated contract-farming-based industrial agroforestry using tripartite and quadpartite model involving industries, farmers, research and financial institutions. Credit facilities are provided to the farmers at the rate of Rs 15,000–20,000 per acre. Industry supplies quality planting material at subsidized rate and assures prevailing marketing price; the farmer assures end-product (timber) to contracting industry, and research institutes advice on site-specific technology. Until now, TNAU has produced and supplied 0.72 million quality planting material to 217 farmers at free of cost, covering 212 ha in five districts of Tamil Nadu using short rotation clones such as Eucalyptus (MTP-1), Casuarina (MTP 1 and 2), *Melia dubia* (MTP 1), Subabool etc.

## Planning Commission

Planning Commission, Government of India, 2001 for promoting agroforestry, has recommended the following:

- Rather than having a uniform strategy for the whole country, commercial agroforestry should be adopted in irrigated districts of the country.
- A separate strategy should be developed for rainfed areas for environmental security, sustainable agriculture (production and economy) and food accessibility.

- Suitable species for commercial agroforestry may include *Acacia nilotica, Bamboo species, Casuarina equisetifolia, Eucalyptus species, Populus deltoides* and *Prosopis cineraria* for different climatic, edaphic and agricultural conditions.
- Specific institutes have been identified for tree improvement and development of clones of specified species.
- Corporate private sector may be encouraged to take up research and development in tree improvement, development of better clones and micro and macro propagation of quality planting material.
- About 100 NGOs may be identified to carry out clonal propagation of seedlings for distribution to farmers at appropriate price and carrying out extension work. Extension activities should include organizing farmers, providing them training in planting techniques, protection measures and other silvicultural operations.
- Technological development to diversify usage of agroforestry species will help to ensure a ready market; for example, bamboo is getting rediscovered as a potential raw material for the development of bamboo composites suitable for use in place of wood and wood composites.
- Bamboo technology mission should be started keeping in view the impending gregarious flowering, followed by mass mortality of bamboo, forest fire famine and insurgency. Circumstances warrant formulation of emergency plans for harvesting and processing of bamboo prior to their flowering.
- As more and more farmers are taking up agroforestry, export-import policies should be modified to encourage agroforestry product marketing.
- A system of market regulation to be put in place, including a mechanism of periodic review in order to protect the interest of both producer and consumer of agroforestry produce.
- A suitable market information system needs to be introduced to inform farmers about major buyers, prevailing prices trends, procedure etc.
- All existing laws executive orders relating to tree felling transport, processing and sale should be amended to facilitate agroforestry.
- Commercial agro forestry may be planned in irrigated districts covering 10 m ha. On annual basis, one million ha should be brought under multipurpose tree species identified by the Task Force. The scheme of NABARD for farm/agro forestry should be expanded and investment of Rs. 100 crores per year should be ensured.
- It is proposed to cover 18 million ha of rainfed areas on watershed basis under agroforestry for conservation of soil and water and plantation of hardy species such as Eucalyptus, Bamboo and Babul. On annual basis 1.8 million ha is proposed for afforestation under various schemes of

Rural Development, NAEB and 'food for work' scheme. An investment of Rs. 2700 crores will be required on yearly basis.

- Major states may establish Agroforestry Cooperative Federation for increasing bargaining powers of farmers in marketing of agroforestry products.
- Wood based industries should continue supply of quality planting material to farmers and ensure suitable buy-back arrangement.

## Conclusion

The policy and scheme stresses that the farmers will be encouraged to take up agroforestry for higher income generation by evolving technology, extension and credit support packages and removing constraints to development of agroforestry. Rural people have been practicing tree planting in their farms and homesteads to meet household requirements of fuel, poles, timber and medicinal plants. With the advent of agroforestry, diversification in agriculture was encouraged to generate high income and minimize risks in cropping enterprises. Agroforestry can be regarded is the only possible alternative at present to meet the target of increasing forest cover to 33% from the present level of less than 25% as envisaged in the National Forest Policy (1988).

## Reference

https://en.wikipedia.org/wiki/BAIF_Development_Research_Foundation

Implementation Guidelines for National Adaptation Fund for Climate Change (NAFCC), Ministry of Environment, Forest and Climate Change, New Delhi, 1–28 pages.

INDC 2015 India. https://www4.unfccc.int/sites/ndcstaging/PublishedDocuments/India%20First/ INDIA%20INDC% 20TO% 20UNFCCC.pdf

Operational Guidelines, National Mission for Sustainable Agriculture (NMSA) 2014 Department of Agriculture, cooperation & Farmers' Welfare, Ministry of Agriculture and Farmers' Welfare, Government of India.

Operational Guidelines, Sub-Mission on Agroforestry (SMAF) 2016 Department of Agriculture, cooperation & Farmers' Welfare, Ministry of Agriculture and Farmers' Welfare, Government of India, 2016.

The story of National Bamboo Mission, Down to Earth, 2018.

Vikaspedia.

Chapter - 4

# Teak based Agroforestry System

## TEAK

Teak (*Tectona grandis* L.) is one of the most important timber trees of India which belongs to the family Verbenaceae, which includes many economically important tropical and subtropical trees. Teak is a tropical hardwood tree species native to India, Burma and Thailand. It is also known as King of timbers, Sagwan, Sag etc. It has small, fragrant white flowers arranged in dense clusters, called panicles, at the end of each branch. These flowers contain both male and female types of reproductive organs (perfect flowers). The large and papery leaves of teak trees are in general hairy on the lower surface. Teak wood, when freshly milled, has a leather-like smell and is highly durable and water resistant. The wood therefore is valued and is used for boat building, exterior construction, veneer, making furniture, carving, turnings, and other small wood projects.

### Botanical Characters

Teak is a large deciduous tree up to 40 m high and 100 cm or more diameter at breast height (dbh). The tree grows straight with grey to greyish-brown branches, and stem is buttressed at the base. The leaves are ovate-elliptic to ovate, opposite about 0.5 m long and 23 cm wide. The branches bear many small white flowers in large, erect, cross-branched panicles. The fruit is a drupe (fleshy, with a stony seed), two-third of an inch in diameter. The bark of the stem is about 1.3 cm thick, and grey or brownish grey in colour. Leaves are 15 to 45 cm (5.9–17.7 in) in length and 8 to 23 cm (3.1–9.1 in) wide, and are held on robust petioles of 2 to 4 cm (0.8–1.6 in) length. Leaf margins are entire. From June to August, fragrant white flowers are borne on 25 to 40 cm (10–16 in) long and around 30 cm (12 in) wide panicles. The corolla tube is 2.5–3 mm long with 2 mm wide obtuse lobes. Fruit setting occur from September to December; fruits are round or spherical with 1.2 to 1.8 cm diameter. Flowers are weakly protandrous, the anthers precede the stigma in maturity and pollens are shed within a few hours of opening of the flower. The flowers are primarily entomophilous (insect pollinated), but can sometimes be anemophilous (wind pollinated). In a 1996 study, it was found that in its native range in Thailand, the bee genus Ceratina were the major pollinator species.

## Wood

Teak, due to a rare combination of superior physical and mechanical properties, is one of the most important timbers in the world. Heartwood is often dull yellowish in colour when freshly cut but turns golden brown or sometimes dark greyish-brown after exposure, often streaked grey or black. Sometimes there are dark patches on it. The wood is oily in touch and there is a leather-like smell in the newly cut wood. Sapwood is whitish to pale yellowish brown. It can easily separate from heart wood. The wood grains are figured well, can be straight, wavy or slightly interlocked, producing an attractive veneer, which is extensively used in the manufacture of furniture and interior fittings.

Growth rings are distinct. Density varies according to moisture content: at 15% moisture content it is 660 $kg/m^{3}$. It dries well but slowly, and the rate of shrinkage is very low. It is very stable in service once dry. Being water resistant, the wood is considered as an excellent timber for building bridge and other construction in contact with water such as docks, quays, piers and floodgates in fresh water. In house building also, teakwood is particularly suitable for interior and exterior joinery (windows, solid panel doors and framing) and is used for floors exposed to light to moderate load. It is also extensively used for making garden furniture and for building beams, transmission line poles, fence posts, wallboards, boxes, musical instruments, toys, railway sleepers and railcar construction.

In other uses, root bark and young leaves produce a yellowish-brown or reddish dye which is used for paper and cloth dyeing and matting. It requires a mordant. The seeds contain oil to the extent of 44.5% which is used in manufacturing of soap.

## Distribution

*Tectona grandis* is considered native to South East Asia; India, Sri Lanka, Indonesia, Myanmar, northern Thailand, and northwestern Laos. The annual rainfall in areas where teak grows averages 1250–1650 mm with a 3–5 month dry season. However, it is found in a variety of habitats and climatic conditions from arid areas with only 500 mm annual precipitation to very moist forests with up to 5,000 mm annual precipitation. In India it is found in Madhya Pradesh, Uttar Pradesh, Gujarat, Maharastra, Kerala, Tamil Nadu, Karnataka. It is introduced in moist deciduous forests of West Bengal, Assam, Bihar, Uttar Pradesh, Orissa, and the Andamans.

## Climate requirement

Teak prefers fairly moist, warm, tropical climate Teak grows and develops well in low-altitude, high-rainfall areas and in climates that have a three to four months dry season. It can withstand extremes of temperature, but maximum temperature of 39–44 °C and a minimum of 13–17 °C is the most preferable. It grows well in rainfall zone of 1200–2500 mm and requires high light intensity.

## Soil

Teak grows well in alluvial soils, with pH range from 6.5 to 7.5. Dry sandy soil, shallow or hard pan soil, acidic, laterite, black cotton and water logged soils are not good for teak and it shows poor growth and form on these soils.

## Varieties

Konni Teak, Godavari Teak, West African teak, South and Central American Teak, Nilambur or Malabar Teak.

## Propagation Technology

**Through Seeds-** The fruits are hard and seed coat needs to be treated to hasten germination. The seeds are pre-treated with hot water, alternate wetting and drying, immersion in cold water for several days, weathering and acid treatment to achieve good germination. The germination percent may vary from 10 to 70. Germination commences in about 15 days, accelerates during the next 15 days, but declines thereafter. Majority of germinates will have appeared in 40 days, when the germinability is around 40%. The beds are watered twice a day for the first two months, once daily for the next three months, and then on alternate days thereafter.

Fresh seeds are difficult to germinate than seeds that has been stored for over twelve months. If it is necessary to use fruits from the same year they were collected, then they should be subjected to alternate wetting then drying for 24 hours each for 14 days. The germination rate is low, usually less than 50%, but sometimes up to 80%. Germination generally increases with increase in size of fruits. Germination of stored seeds can be enhanced by soaking them before sowing for 24–48 hours in warm water, and changing the water frequently. Another method of encouraging germination is to char (or half burn) the fruits by covering them with a thin layer of grass and lighting it. Germination usually starts 10 days after sowing but the seeds may take as much as 2–3 months to germinate. Seedlings should be given a little shade in the first year and after that they can be planted into their permanent positions in fields. Natural regeneration particularly occurs in patches in forests exposed to fires. Seeds are generally collected from the forest floor to establish plantations. Seeds of 20 years old trees are recommended and are often collected from selected stands. Fruit that has been dormant in the ground for 30–40 years has been known to germinate abundantly.

Among vegetative propagation methods, both grafting and budding methods shows better results than branch cutting methods. The rooting time of cutting is 8–15 days, and the survival rate of rooted stock is 90 -100%.

**Through Stumps-** Generally stumps or seedlings are used as planting material. The seedlings have to be maintained in the nursery for about one year for stump preparation. Then the seedlings are uprooted, all the leaves and secondary roots are removed except 2.5 cm, shoot with 22.5 cm tap root portion. The stem portion receives an oblique cut and the root portion a horizontal cut. All laterals from the tap root are pruned away. The stumps should preferably be planted within 2 to 3 days of pruning.

### Planting

The land should be ploughed thoroughly and pits of size 45 cm × 45 cm × 45 cm are prepared. The planting spacing should be 2 m × 2 m or 3 m × 3 m or 3 m × 4 m. Pits are filled with a mixture of farmyard manure and soil. Seedlings should be planted in the pits during rainy season. For stump planting, however, pitting is not required and crowbar may be used.

### Irrigation

In monsoon month no irrigation is required. Irrigation should be applied in hot or summer month or during stress periods. Irrigation during stress period improves yield to a great extent. Drip irrigation is beneficial in farm lands. Irrigation reduced the rotation period and also enhances the productivity.

### Thinning

First thinning is done at 5–10 years after planting depending upon site quality and initial spacing. Generally under good site and close spacings, first and second thinning is done at 5 and 10 years, leaving 25% of the trees for further growth and development. The spacing of trees and the frequency, timing and intensity of thinning strongly affect the pattern of growth and yield in a plantation. By increasing the frequency of irrigation and thinning, diameter growth can be increased.

### Nutrient management

N:P:K (15:15:15) @ 50 gm per plant should be applied in August and September months for first three years.

**Table 4.1:** Plant Protection of Teak

| Pest | Control |
|---|---|
| Teak defoliator, skeletonize are serious pest of teak and they damaged tree to greater extent. | To get rid of this pest spray with quinalphos@ 300ml/ 150 L of water. |
| **Disease** | **Control** |
| Pink disease, powdery mildew and root rot are major disease of teak plant. | It can be controlled by spraying of M-45@ 1 Kg/ha. |

### Yield

Each teak tree yields about 0.3 to 0.4 cubic m wood in 14 years. The average productivity of teak in teak plantations in Nilambur, Kerala was 2.85 cubic m/ha/ year in 53 years rotation period. Recent studies conducted on teak growing in farmlands with irrigation, fertilizer application and management revealed the possibility of reducing the rotation period to 25 years and also achieve increased productivity. The trees in farmlands grow faster and produce more biomass than those in plantations in the forest areas. The quality of teak timber in 12 years old plantation was found to be similar to that of 20 years old tress of forest land. A tree can produce a maximum of 0.60 cubic m of timber in 20 years under best management and there could be a maximum of 100 trees per acre (KFRI, 1979).

## Utilization

Teak wood is globally valued for its strength, durability, dimensional stability, working quality and noncorrosive property when coming in contact with metal. The durability is attributed to the deposition of polyphenols in its heartwood. On account of these outstanding properties, teak is sometimes hailed as the King of timbers. It is mainly used in furniture, cabinet making, various grades of plywood, panelling, construction, poles, piles, ship building and other purposes. Teak wood is being increasingly used by the plywood industry in large quantities for high grade commercial and tea-chest categories of plywood. Lops and tops and other rejects of the manufacturing process are used as fuel wood. The seeds contain oil to the extent of 44. 5% and the oil is used in soap manufacture.

## Teak in Agroforestry Systems

**Fig. 4.1:** Teak based agroforestry system (BAU, Ranchi, Jharkhand)

Teak is one of the most liked or preferred silvicultural species by the farmers. It is planted in different models, combinations as well as in different spacings. IFGTB (Institute of Forest Genetics and Tree Breeding) has developed agroforestry models like agri-silviculture (teak+casuarinas with agricultural crops such as maize, cotton, turmeric, tomato and chilly), agri-silvi-horticulture (teak+coconut with agricultural crops such as ginger, turmeric, vegetables, maize and cotton) and silvi-horticulture model (teak-guava, annona). Under irrigated lands, silvi-pasture model was developed with teak and casuarina as the tree component and napier and guinea as pasture components. Teak as the most suitable tree species for adopting Teak based agroforestry system of land use in hill zone of Karnataka. Total income from field crops + saptoa + Teak was 46 percent higher than growing field crops + sapota.

## Economics of Teak based agroforestry system

Teak farming is highly profitable farming as it has great demand in India as well as abroad. The economics of teak based agroforestry system gave higher gross income, net return, B:C ratio than open farming system (sole crops and guava tree alone).

By planting Teak approximatily 0.30 to 0.40 $m^3$ wood from one teak tree is obtained. The average value of teak is approximately Rs. 88000 per cubic metre in Indian timber market. So, it will earn around 30000/teak tree. If 833 (4 m × 3 m spacing) trees are planted in a ha it will earn 833 × 30000= Rs. 2,49,90,000/ha from a 25 years old well managed plantation.

**Table 4.2:** B:C ratio analysis of different teak based agroforestry system in India

| Agro-forestry system | B/C ratio | References |
|---|---|---|
| Teak + *Ocimum tenuiflorum* | 1.01 | Kumar, *et al.*, 2015 |
| Teak + *Ocimum gratissimum* | 1.24 | |
| Teak + *Ocimum basilicum* | 1.13 | |
| Teak + Paddy | 2.20 | Barai, T., 2016 |
| Teak + Arhar | 2.43 | |
| Teak + Colocasia | 1.43 | |
| Teak + Ginger | 1.53 | |
| Teak + Okra | 2.85 | |
| Teak + Brinjal | 2.38 | |
| Teak + Chilli | 2.98 | |
| Sole Teak + Patchouli | 1.94 | Kumar *et al.*, 2016 |
| Sole Patchouli100% FYM (20 t/ha ) | 1.87 | |
| Teak + Patchouli100% FYM (20 t/ha ) | 2.14 | |
| Sole Patchouli100% VC (10 t/ha ) | 2.07 | |
| Teak + Patchouli100% VC (10 t/ha ) | 2.32 | |
| Sole Patchouli100% N C (0.5–1.0 t/ha ) | 1.87 | |
| Teak + Patchouli100% N C (0.5–1.0 t/ha ) | 2.04 | |

| Agro-forestry system | B/C ratio | References |
|---|---|---|
| Sole Patchouli 50% FYM + 50% NC | 1.79 | |
| Teak + Patchouli 50% FYM + 50% NC | 2.10 | |
| Sole Patchouli 50% VC + 50% NC | 1.91 | |
| Teak + Patchouli 50% VC + 50% NC | 2.21 | |
| Sole Patchouli 50% FYM + 50% VC | 2.01 | |
| Teak + Patchouli 50% FYM + 50% VC | 2.35 | |
| Sole Rice | 2.24 | Verma, V., 2016 |
| Teak + Rice | 2.36 | |
| Sole Sugarcane | 1.36 | Mali *et.al.*, 2017 |
| Teak+ Sugarcane | 3.90 | |
| Sole Teak-8 × 2 × 2m | 0.00 | Deshmukh, P., 2018 |
| Teak+cowpea - Fodder sorghum | 2.42 | |
| Teak+cowpea -Fodder Maize | 2.39 | |
| Teak+Blackgram -Fodder sorghum | 2.39 | |
| Teak+Blackgram -Fodder Maize | 2.38 | |
| Teak + cowpea + mulching+Fodder Maize | 2.42 | |
| Teak+Blackgram + mulching+Fodder Maize | 2.43 | |
| Teak + Pineapple + Cowpea-Potato-Okra | 2.37 | Mohapatra, S. C. and Behera, S., 2018 |
| Teak + Pineapple + Okra-Frenchbean-Brinjal | 2.76 | |
| Teak + Pineapple + Brinjal-Cowpea-Amaranth | 2.69 | |
| Teak + Papaya + Cowpea-Potato-Okra | 1.27 | |
| Teak + Papaya + Okra-Frenchbean-Brinjal | 1.37 | |
| Teak + Papaya + Brinjal-Cowpea-Amaranth | 1.57 | |
| Sapota + Teak + Field crop | 3.01 | Mutanal, S. M. and Hundekar, S. T., 2018 |
| Sole Field crop | 2.21 | |

## Conclusion

Teak, being the most valuable and widely produced tropical hardwood species, is known as king of timbers and national and international demand for teak timber exceeds the yield from natural forest and plantations. This high demand creates opportunities for farmers and teak is now grown in agroforestry system in many countries. These systems, in addition to meeting commercial needs for timber and improving environmental conditions, also enable farmers to diversify farm production, reducing risk associated with traditional farming with field crops only, and generate much needed income with high monetary returns and higher benefit cost ratio.

## References

Barai T 2016 Performance of different Teak based agroforestry system in Cuttack District of Odisha. M. Sc. thesis, Orissa University of agriculture and Technology, Bhubaneswar, Odisha, 51 page.

Chaturvedi AN and Khanna LS 1994 Forest Mensuration (2nd ed.). International Book Distributors, Dehradun, India.

Deshmukh P 2018. Economics as influenced by planting geometry Rabi fodder @ Rahuri. Paper presented in annual group meeting of AICRP- Agroforestry, 28–30 July 2018, Birsa Agricultural University, Ranchi, Jharkhand.

KFRI 1979 Yields from teak plantations in Kerala. Kerala Forest Research Institute.

Kumar M, Thakur NS and Hegde HT 2015 Growth, herb yield and financial flows from *Ocimum spp.* Intercroped under teak (*Tectona grandis* L. f.)-*Ocimum spp.* based silvi-medicinal system in Gujarat, India. *International Journal of Innovative Horticulture* **4**(2) 113–118.

Kumar D, Bijalwan A, Kalra A. and Dobriyal MJR 2016 Effect of shade and organic manure on growth and yield of patchouli [*Pogostemon cablin* (blanco) benth.] under teak (*Tectona grandis* L. F.) based agroforestry system. *Indian Forester* **142**(11) 1121–1129.

Mali SC, Shedage S and Shrivastava PK 2017 Economic Evaluation of Sugarcane Based Agro forestry Systems. *J tree Sci.* **36**(1) 34–37.

Mohapatra SC and Behera S 2018 Crop yield and economics of agrisilvicultural system. Paper presented in annual group meeting of AICRP- Agroforestry, 28–30 July 2018, Birsa Agricultural University, Ranchi, Jharkhand.

Mutanal SM and Hundekar ST 2018 Economic Evaluation of Sapota-Timber Tree Species. Paper presented in annual group meeting of AICRP- Agroforestry, 28–30 July 2018, Birsa Agricultural University, Ranchi, Jharkhand.

Tiwari DN 1992 A monograph on teak (*Tectona grandis*) Linn.f. International Book Distributors, Dehradun, India.

Verma V 2016 Performance of different tree species on the rice field bunds. M.Sc. thesis, Orissa University of agriculture and Technology, Bhubaneswar, Odisha, 43 page.

Chapter - 5

# Gamhar based Agroforestry System

## GAMHAR

*Gmelina arborea,* commonly known as White Teak or Gambhar or Gumhar or Sewan, is an indigenous tree species of India which belongs to the family of Verbenaceae. *Gmelina arborea* is fast growing tree and its wood is used for timber, furniture wood, paper and pulp making and other forest-based industries. It is one of the most promising multipurpose tree species which grows up to 30 m in height, has diameter of over 80 cm and is able to fulfill the need of fuel, fodder, wood, medicinal products, other domestic needs etc. It is very light-demanding and is an ideal choice for farm forestry and large-scale afforestation/ reforestation initiatives. It is widely grown as a component of agroforestry system in humid tropics and contributes to 0.47% of total volume of trees under Agroforestry systems the country (ISFR, 2013).

### Botanical Characters

*Gmelina arborea* is one of the fast growing indigenous multipurpose tree species which produces one of the best quality timbers in India. It is a medium- to large-sized deciduous tree of up to 30 m height. The tree form is fair to good, with 6–9 m of branchless, often crooked trunk, and a large low-branched crown. It has a smooth whitish grey (ashy) in colour and corky bark, have warts with lenticular tubercles exfoliating in regular patches when grow old. Leaves are simple, opposite, ovate-deltoid, entire, coriaceous, densely tomentose below, 3-nerved from the base, acuminate at apex, cordate at base. Inflorescence is axillary or terminal panicles. Flowers are brownish yellow; have cup shaped calyx, 5-toothed, teeth very small or obsolete, externally tomentose; have tubular corolla, 2-lipped, ventricose, externally tomentose; with 4 stamens, included in the corolla tube.

### Distribution

The species is found mainly in Bangladesh, Cambodia, China, India, Laos, Myanmar, Nepal, Pakistan, Sri Lanka, Thailand, and Vietnam. In India, it is distributed along eastern sub-Himalayan tract, Indo-Gangetic plains, Aravali Hills, central India, western Peninsula and western Himalayas. It grows well in eastern sub-Himalayan belt but also extends into mixed-deciduous forests of central India and sal forests of Uttar Pradesh, Bihar and Orissa, occasionally found in wet evergreen forests of the Western Ghats. In the north-west Himalayas it ascends to 1000 m in valleys and over ridges (Gupta, 1993).

## Climate requirement

The species occurs in a variety of forest habitats, including tropical semi-evergreen, submontane, very moist teak forests, deciduous sal, and dry teak forests. It occurs naturally from 58 to 30°N latitude between 50 and 1300 m elevation in areas with distinct dry seasons. It is a hardy plant and can be grown in both tropical and subtropical conditions. It likes humid climate with the optimum temperature range from 20 to 38 °C. The annual rainfall requirement is 750–4500 mm. It grows very well in high-sunshine, low-shade areas.

## Soil

*Gmelina arborea* is well adapted to a wide range of soils. It grows very well in red sandy loam fairly deep soils with a pH of 5 to 8 but can also be grown on other soils.

## Planting stock

Tree improvement programme for *Gmelina arborea* in India was started during early 1990s. Improved accessions and seeds from the clonal seed orchard are available in Rain Forest Research Institute, Jorhat. The seeds from superior candidate plus trees (CPTs) are available at Institute of Forest Genetics and Tree Breeding (IFGTB).

## Propagation Technology

*Gmelina arborea* is propagated by both seeds and vegetative methods.

i. **Through Seeds**– Seeds are sown directly in the root trainer or on raised beds of 0.6 m height and 1m broad within convenient length. The beds are filled with sand up to the height of 0.5 m. The seeds are closely placed in rows 5 cm apart. Seed beds should be watered twice a day with the help of rose cane. There are 700–1400 seeds in one kilogram (Evans, 1982). Germination percentage can vary from 60 to 85.

ii. **Clonal propagation method**– It is one of the most convenient, less expensive and successful method specially for multiplication of selected superior CPTs. The cuttings are collected in the early morning between 6–9 am and after treating with 750 ppm of Indole -3 Butyric Acid (IBA) the cuttings are immediately planted in the root trainer filled with the vermiculite. The root trainers are placed inside the low cost poly tunnel or mist chamber. The internal temperature and humidity should be maintained at 26–36 °C and 80–100% respectively. The dried and dead cuttings are frequently removed to avoid the spread of fungus. Roots of cuttings are generally observed in 15–25 days after planting. About 3 months old cuttings are transplanted in clonal plantations.

iii. **Stump Planting**– For stump planting, 90 seeds per square meter should be sown. Seedlings are usually ready for stump preparation in 7–8 months and should have a minimum root collar diameter of 2.5 cm. The stem and roots of seedlings should be pruned to 5 cm and 20 cm, respectively, before planting.

## Planting

Pits of 45 cm × 45 cm × 45 cm are dug in May and kept open for 15 days. They are then filled with standard potting media (2:1:1, soil : FYM : sand). Monsoon planting (June–Aug) is ideal as the seedlings get properly established in field conditions. The right age of seedlings suitable for planting is 6–7 months. Square and rectangular planting of *G. arborea* is preferred in agroforestry (6 × 6 $m^2$, 4 × 4 $m^2$, 4.5 × 4.5 $m^2$ spacing). For the production of sawn wood, planting is done at 3 m × 3 m spacing, whereas for producing raw material for the pulp and paper industry, planting is done at 1.2 × 1.2 $m^2$ or 1.8 × 1.8 $m^2$.

## Yield

Based on the purpose, plantation-grown trees can be harvested between 8 and 15 years. The growth of the species is remarkably fast and on good sites the tree can reach 20 m height in 5 years. The tree attains more than 30 m height with about 60 cm dbh at maturity. Form of the tree is fair to good, with 6–9 m of clear bole. *G. arborea* trees are harvested for pulp and firewood after 4–5 years of planting, and 10–12 years for log production. Under a good management regime each tree yields about 1.5 to 2 tons/plant so per hectare yield could be in the range of 250–300 tons/ha. The wood of *G. arborea* fetches Rs.8000/tone in local market.

## Utilization

The wood takes nails, paint and stain well, and because of smooth grain and light colour of its wood, it is referred to as 'White Teak'. The wood has moderately low density of 380–430 kg/cu m when harvested at approximately 8 years of age, at most locations. Its calorific value is about 4400–4800 kcal/kg. Relatively good uniformity, stability and light colour make its wood a suitable raw material for a number of products including pulp, paper, medium density fibre board, laminated veneer lumber, wood cement composites, particle board, furniture, furniture interiors, plywood interiors, finger jointed lumber, door panels, pallets, pencils, and match sticks. It is also used for making good quality toys and picture frames.

## Gamhar in Agroforestry System

The tree has immense potential for its timber as well as its medicinal value. It is a preferred species by farmers, forest departments and ayurvedic industries due to the multipurpose utility, rapid growth, and high economic returns. *G. arborea* is grown as an agroforestry tree in coffee and cocoa plantations to protect the young plantations. Rice, maize, beans, black gram, cowpea, cassava, peanuts etc. are grown along with this species as intercrops. Growing *G. arborea* in agroforestry models is a commercial practice in Tamil Nadu. Some of the common models are Gmelina + Groundnut, Gmelina+ Watermelon, Gmelin a+ Pulses, Gmelina + Maize, Gmelina

+ Banana. Multitier cropping system also has been followed in Pudukkottai district of Tamil Nadu, for example, Coconut + Gmelina + Banana + Pepper (pepper is trained on the gamhar trees).

**Fig. 5.1:** Gamhar based agroforestry system (BAU, Ranchi, Jharkhand)

## Economics of Gamhar based agroforestry system

The study was conducted with Gamhar (*Gmelina arborea*) and Mango (*Mangifera indica*) as silvi and fruit trees respectively intercropped with five arable crops namely Pigeon pea, Ground nut, Okra, Bottle gourd and Maize in *kharif* and Mustard in *rabi* season. The gross income under each treatment included income from different intercrops. The results showed that the integration of crops with components of tree and fruit gave higher gross income than tree and fruit tree alone. The combination of gamhar + mango – okra – mustard was found to be superior with maximum gross income of Rs. 1.645 lakh followed by gamhar + mango-ground nut-mustard, Rs. 1.504 lakh (Momin, *et al.*, 2017).

Kumar, S. (2013) reported that highest financial yield of gamhar (Rs. 72,972.00 of total boundary plants), papaya (Rs. 9,40,287.00 per ha), pea (Rs. 29,280.00 per ha), gram (Rs. 61,875.00 per ha), and Indian mustard (Rs. 15,540.00 per ha) in gamhar based agroforestry system as compared to financial yield of gamhar (Rs. 21,406.00 of total boundary plants), papaya (Rs. 8,10,101.00 per ha), pea (Rs. 28,120.00 per ha), gram (Rs. 51,095.00 per ha), and Indian mustard (Rs. 14,595.00 per ha) in sole plantation.

**Table 5.1:** B:C ratio analysis of different Gamhar based agroforestry system in India

| Agro-forestry system | B/C ratio | References |
|---|---|---|
| Gamhar + Pineapple | 1.65 | Nayak *et al.*, 2014 |
| Gamhar + Aloevera | 1.30 | |
| Gamhar + Kalmegh | 1.56 | |
| Gamhar + Mangoginger | 1.37 | |
| Sole Pineapple | 1.58 | |
| Sole Aloevera | 1.16 | |
| Sole Kalmegh | 2.27 | |
| Sole Mangoginger | 2.19 | |
| Gamhar + Mango + Pigeon pea | 3.20 | Momin, *et al.*, 2017 |
| Gamhar + Mango + Black gram | 3.30 | |
| Gamhar + Mango + Lady's finger | 2.80 | |
| Gamhar + Mango + Bottle gourd | 3.60 | |
| Gamhar + Mango + Maize | 3.70 | |
| Gamhar + Mango | 1.90 | |
| Sole Gamhar | 1.80 | |
| *G.arborea* + Mango + Arhar | 1.18 | Annual Report, 2017–18 |
| *G.arborea* + Mango + Cowpea – Toria | 1.52 | |
| *G.arborea* + Mango + Greengram – Toria | 1.49 | |
| Sole *G.arborea* + Mango | – | |
| Sole *G.arborea* | – | |
| Sole Mango | – | |
| Sole Arhar | 1.01 | |
| Sole Cowpea –Toria | 1.49 | |
| Sole Greengram – Toria | 1.44 | |
| *G.arborea* + Sweet orange + Arhar | 1.18 | Annual Report, 2017–18 |
| *G.arborea* + Sweet orange + Cowpea – Toria | 1.52 | |
| *G.arborea* + Sweet orange + Greengram – Toria | 1.49 | |
| Sole *G.arborea* + Sweet orange | – | |
| Sole *G.arborea* | – | |

| Agro-forestry system | B/C ratio | References |
|---|---|---|
| Sole Sweet orange | – | |
| Sole Arhar | 1.11 | |
| Sole Cowpea –Toria | 1.45 | |
| Sole Greengram – Toria | 1.42 | |
| *Gmelina arborea*+ *Cajanus cajan* | 2.05 | Berry, N., 2018 |
| *Gmelina arborea*+ *Withania somenifera* | 3.80 | |
| *Gmelina arborea*+ *Rauvolfia serpentina* | 6.42 | |
| Sole *Gmelina arborea* | 2.03 | |
| Sole *Cajanus cajan* | 0.72 | |
| Sole *Withania somenifera* | 0.90 | |
| Sole *Rauvolfia serpentina* | 8.09 | |
| *G.arborea* + Arhar | 2.94 | Ghode B. D. and Dongre, R., 2018 |
| *G.arborea* + Cowpea–Toria | 3.68 | |
| *G.arborea*+ Greengram–Toria | 4.01 | |
| Sole *G.arborea* | – | |
| Sole Arhar | 2.96 | |
| Sole Cowpea-Toria | 4.30 | |
| Sole Greengram-Toria | 4.61 | |

## Conclusion

The results pointed out to the immediate need for development of agriculture in rainfed areas together with gamhar trees. Crops like cereals, pulses, oilseed; vegetable crops and medicinal and aromatic plant grown as intercrops demonstrated the potential for attaining maximum and sustainable gross monetary returns under erosion prone areas, where farmers are predominantly marginal and resource poor.

## References

Annual Report 2017–18 Yield and economics of *Gmelina arborea* + Mango based agroforestry system, Regional Research Station (Red & Laterite Zone), Bidhan Chandra Krishi Viswavidyalaya, Jhargram, West Bengal.

Annual Report 2017–2018 Yield and economics of *Gmelina arborea* + Sweet orange based agroforestry system, Regional Research Station (Red & Laterite Zone), Bidhan Chandra Krishi Viswavidyalaya, Jhargram, West Bengal.

Berry N 2018 Economic analysis of Gmelina based silvi-agri-medicinal system. Paper presented in annual group meeting of AICRP- Agroforestry, 28–30 July 2018, Birsa Agricultural University, Ranchi, Jharkhand.

Evans J 1982 Plantation forestry in the tropics. Clarendon Press, Oxford, UK. 472 page.

Ghode BD and Dongre R 2018 Yield and economics of *Gmelina arborea* based agrisilvicultural system. Paper presented in annual group meeting of AICRP–Agroforestry, 28–30 July 2018, Birsa Agricultural University, Ranchi, Jharkhand.

Gupta RK 1993 Multipurpose trees for Agroforestry and Wasteland Utilization.Oxford and IBH Publishing Co. Pvt. Ltd. New Delhi. 557 page.

KFRI 2002 Plantation Technology for Nine Indigenous Tree Species of Kerala. KFRI (Kerala Forest Research Institute), Research Report 231.

Kumar S 2013 Studies on growth performance and economic evaluation of gamhar (*Gmelina arborea* roxb.) based agroforestry system in East Singhbhum District in Jharkhand. M.Sc. thesis, Birsa Agricultural University, Kanke, Ranchi, Jharkhand, 1-77 pages.

Momin BG, Dhara PK and Tarafdar PK 2017 Differential responses of arable crops with gamhar (*Gmelina arborea*) and mango (*Mangifera indica*) based agroforestry systemin red and lateritic soils of West Bengal, India. *Indian Journal of Agricultural Research* (51) 86–89.

Nayak MR, Behera LK, Mishra PJ and Bhola N 2014 Economics and yield performance of some short duration fruit and medicinal crops under agri-silvicultural system in rainfed uplands of Odisha. *J. Appl. and Nat. Sci.* **6**(1) 274–278.

## CHAPTER - 6

# Poplar based Agroforestry System

## POPLAR

*Populus deltoides* is a deciduous tree, belongs to the family Salicaceae and native of North America is economically the most important species being grown under Agro-forestry in north India, commonly known as aspen, poplar and cotton wood. Poplar wood and bark are used for making plywood, boards, sports goods, and also for making match-sticks and pencil. In Northern India, exotic species such as *Populus* is being grown on an increasing scale as trees of agroforestry by farmers on field bunds, within agricultural fields, wastelands etc. The popularity of these species is mainly because they are fast growing, tolerant to pruning and have small canopy which casts minimum shade effects. It occurs inter spread throughout the forest of the temperate and cold regions of the northern hemisphere. Since its introduction in India in the year 1981, poplar plantation started mushrooming in Northern part of India and farmers are taking various crops like sugarcane, wheat, maize etc as intercrop with poplar. Poplar is widely planted above $28^0$ N latitude in India in the states of Arunachal Pradesh, Haryana, Himanchal Pradesh, Jammu & Kashmir, North Bengal, Punjab, Uttar Pradesh, and Uttarakhand. Area under poplar and willow is estimated to be 3, 62,700 ha in India. The largest area of 3,12,000 ha is covered by *Populus deltoides* (poplar) plantations outside the forest. The annual returns from its cultivation are estimated to be around one lakh rupees per acre per year (National Poplar Commission of India– Country Report (2012): Poplars and Willows in India).

### Botanical Characters

*Populus deltoides* is fast-growing, medium to large in size, 20–30 m tall having 100 cm dbh (diameter at breast height). Bark is greyish-green and smooth at first, later blackish and furrowed; trunk short and massive in the open, often divided into a few large, wide spreading limbs near the ground to form a broad, irregular-shaped open crown. Leaves are broadly deltoid, 8–15 cm long with nearly same breadth, glabrous on both sides, short– acuminate, dentate with incurved glandular or callous tipped teeth; bases truncate to subcordate with 2–3 basal glands; petiole strongly flattened laterally. Bracts of catkins fringed or fimbriate, the divisions narrow; staminate aments 7.5–12.5 cm long, thick; stamens about 60 or more; anthers red; pistillate aments green and slender; ovaries glabrous; stigmas 3 or 4.

Mature seed catkins are 15–25 cm long; stalk hairless; capsules ovoid, 6–10 mm long, glabrous, green, splitting into 3–4 parts when mature; peduncle 3 mm long; seeds cottony. *P. deltoides* is dioecious with sex ratio of 1:1. Male buds are much larger buds and develop somewhat earlier than female buds. Flowering occurs before leaves appear.

## Distribution

Among various species of *Populus, P. deltoides* is the most widely planted species in India. It was introduced in India in the late 1950s. It is planted in plains of North–West India, *i.e.*, Western Uttar Pradesh, Punjab and Haryana and to some extent in the outer plains/valleys of Uttarakhand and Himachal Pradesh. It is widely planted in the Tarai belt extending up to Bihar and West Bengal.

## Climatic requirement

These trees grow mostly in temperate climatic conditions. In India, poplar trees thrive best in areas with temperature between 5–45 °C.

## Soil

It grows best on fertile loamy soils rich in organic matter. Alkaline and saline soils should be avoided for poplar cultivation. Ideal soil pH is between 5.8 to 8.5.

## Planting material/ Propagation Technology

Poplar is mainly propagated by cuttings but can also be propagated through seeds.

**Through Seeds**– The spikes ripen during June–July. Branches bearing spikes are collected by cutting shortly before the seed is about to shed. The capsules are dried for two to three days. Seeds are then collected with the help of cotton spun and then picked manually. Poplar seeds remain viable for about three weeks at room temperature. Seedlings are raised in polypots, kept under shade and watered twice daily. Seedlings grow very fast. Raising of plants from seeds is generally practiced for tree improvement.

**Through Cuttings**– Poplars are generally raised by cuttings. The optimum diameter of cutting varies from 1–3 cm and length around 20–22 cm. Both ends of the cuttings should be sealed by wax to protect from moisture loss. The cuttings should be submerged in fresh water for 48 hours immediately after cutting. Prior to planting, cuttings should be drenched with Aldrin (30 EC thoroughly mixed in 100 litres water) emulsion. Then the cuttings are inserted vertically in well prepared nursery beds. The spacing between cuttings is generally kept at 50–60 cm within the row and 60–80 cm between the rows. The entire length of cutting should be inserted into the soil keeping only one bud above ground level. The nursery raised plants are called entire transplants (ETPs). ETPs attain a height of about 3–4 m in single growing season and are utilized for planting in the field.

## Time of planting

January and February is the best time for transplantation of poplar seedling. However, planting can be done up to 10 March.

## Spacing

Spacing of 5 m × 5 m (400 plants/ha), 5 m × 4 m (500 plants/ha), 6 m × 2 m (833 plants/ha) or 5 m × 2 m (1000 plants/ha) can be followed as per the purpose of plantation. Wider spacing is followed if agroforestry is to be practiced and lesser spacing in case of sole plantation. Spacing also varies with type of wood requirement.

## Seedling Treatment

Seedlings should be treated with fungicide (Emisan 6 @ 200gm/100 L for 20 minutes) before planting to protect them from fungal diseases. Thick and damaged roots of seedlings should be pruned before treatment. Roots of seedlings are then dipped in a solution of Chlorpyriphos @ 250ml/100 L of water for 10 to 15 min for termites and other pests.

## Transplanting

Pits of about one meter depth and diameter of 15 cm are dug with the help of soil augers. The ETPs (Entire Transplants) are then put vertically into the pit and the pit is then filled with the mixture of top soil, 5 kg FYM, 25g MOP, 200g SSP, 25 g micronutrients (zinc, iron, copper, etc.) and *Trichoderma*. One month prior to transplanting, FYM is mixed with *Trichoderma* and kept it in shade in such a way that the moisture is not lost and the mixture doesn't dry.

## Irrigation

In nursery, apply irrigation immediately after planting. Afterwards irrigation should be applied at 7 to 10 days intervals depending upon soil type and climate. In the main field, light and frequent irrigation with interval of 7 days should be applied till the onset of monsoon during first year. In winter months (October–December) two irrigation per month should be applied. During second year, irrigations should be applied at an interval of 15–20 days in winter season and 7 days in summer. Third year onwards, two irrigations per month in summers should be applied.

**Table 6.1:** Fertilizer and manures requirement (per plant) of Poplar

| Age of crop (Year) | FYM (kg) | Urea + SSP(gm) |
|---|---|---|
| First year | 8 | 50 |
| Second year | 10 | 80 |
| Third year | 15 | 150 |
| Fourth year and above | 15 | 200 |

Fertilizer should be applied in the month of June, July and August every Year.

**Table 6.2:** Plant Protection of Poplar

| Pest | Control |
|---|---|
| Termites: | If infestation is observed, to control apply Chlorpyriphos @3.75 L/ha. |

| Pest | Control |
|---|---|
| Stem Borer: If infestation is observed, | Control apply Phorate 10G @ 12.5 kg/ha upto second year of plantation or inject Kerosene @ 2 to 5ml per hole. |
| Mites: | If infestation is observed, to control spray Metasystox @ 2ml/L of water. Repeat the spray with interval of 15 days. |
| Leaf Defoliators: Its infestation is observed in July month. | To control spray Chlorpyriphos 500 ml + Cypermethrin @ 200 ml/ha. |

| Disease | Control |
|---|---|
| Collar rot: | As preventive measure, treat seedling root with Emisan 6 @ 4 to 5 gm/plant. |
| Blight: | Its infestation is seen in the month of August and September. If infestation is observed, to control spray crop with Carbendazim @ 2gm/L of water. |
| Wilt: | Infestation is observed in the month of May and June. To control spray wettable Sulphur powder @ 1.25 Kg/ha. |

### Variety

- G 48: Most popularly grown in plains of Punjab, Haryana and UP.
- W 22: Suitable for cultivation in Himachal Pradesh, Pathankot and Jammu.
- Udai, A 26, C 8, C 15, S 7, W 32, W 39.

### Yield

Poplars can produce moderate quality veneer logs at short rotation period of 6 to 8 years, achieving a high productivity of 20–25 cubic metre of wood/ha/yr. In India, poplar is usually harvested when it attains dbh (diameter at breast height) of 24 cm. The productivity of well-managed poplar plantations in Punjab is estimated to be 46.92 cubic metre of wood/ha/yr, whereas the average productivity of natural forests in India is only 0.7 cubic metre of wood/ha/yr (Verma, 1993).

### Utilization

Poplars were initially introduced in India to meet the needs of match industry. Later, poplar wood was started to be used extensively in making plywood. Poplar veneers are not suitable for classification as Type A surface (Bansal, 1999). Hence, its veneers are used only as core and cross band veneers in plywood industry. The earliest contribution of poplar trees to the farmers is as fuelwood from pruning. It is also used in packing cases for fruits and other food stuffs. It is also an excellent source of fibre for various grades of paper (fine paper, packing paper and newsprint).

**Intercropping:** Intercropping is always preferred as it provides agricultural returns on the one hand and increased growth of poplar on the other due to frequent irrigation and hoeing of agricultural crops. Intercropping can be done in the first two years of poplar plantation with close spacing. It helps farmers to get income during initial years of poplar cultivation when tree is not giving any return. Crops like sugarcane, turmeric and ginger can be taken as intercrops. However, agroforestry can be practiced successfully in widely spaced poplar plantation for many years.

## Poplar in Agroforestry Systems

Cultivation of poplar under agroforestry systems by farmers started in the 1970s. Initially poplar under agroforestry systems was grown only by rich and progressive farmers who had surplus land and financial capacity to take the risk of the plantation failure. Since then, poplar based agroforestry system is considered to be economically viable and more sustainable than many other crop rotations prevalent in northern and eastern India (Singh and Sharma, 2007). Its popularity among farmers is mainly due to multi-utility wood, fast growth, high market prices, less competition with associated crops and pruning-tolerant nature (Chauhan *et al.* 2012). The poplars, having straight bole, shedding leaves during winter month thus allowing enough light for companion crops and combining well with the agriculture crops, giving good economic returns in short rotation *i.e.* 5–7 years and easy availability of bank loans, have become the most popular tree species for planting under agro-forestry systems after 1980s. The credit for taking poplar among farmers for widespread plantation gocs to WIMCO (Western India Match Company), that started an extension project for massive plantation of poplar in Punjab, Haryana and western Uttar Pradesh with the financial support of NABARD (National Bank for Agriculture and Rural Development). The poplar scheme of WIMCO–NABARD was a noble venture of industry and banking sector partnership for the benefit of the industry, farming community, and the unemployed, besides providing environmental services.

Poplar has become the most preferred tree species for commercial agroforestry systems in north–western states like Punjab, Haryana, Uttar Pradesh, Himachal Pradesh and Arunachal Pradesh. *P. deltoides* is the only species of polar that is planted on a significant scale in India. It constitutes the backbone of agroforestry in irrigated plains of Northern India. Clones G–48, Udai, W–22, W–32, W–39, A–26, S–7 C–15, S–7 C–8 are especially suitable for agroforestry. Poplars are raised on farmlands in rows, along boundaries, paths, and in blocks. In the terai belt these are intercropped at 5 m × 5 m spacing which easily allows tractor operations. *Kharif* as well as *rabi* crops can be grown in the field throughout the rotation of poplar when poplar is planted on farm boundaries. Agricultural crops such as sugarcane, potato, berseem, paddy, wheat, maize, gram, bean, pea, lentil, lobia, jowar, oats, soybean, mustard, ginger, turmeric, vegetable crops like tomato, chillies, garlic, raddish, colocasia, potato, cabbage, spinach, etc. and fruit crops such as citrus, guava, mango, etc. can be profitably grown with poplar. can be grown as intercrops. During rabi, because of its deciduous character, allow winter crops only with marginal effect on crop yields.

Poplar-wheat and poplar-mustard are the most common combinations under agri-silviculture systems in Punjab. However, in block plantation of poplar, the usual *kharif* crops can be grown for two years only; thereafter shade loving crops like ginger, turmeric, etc. can be taken. Considering the fast rate of growth, ease of establishment, easy marketing and multifarious uses, poplars have grown under various systems *viz.* Agri-silvicultural, Agri-silvi-horticultural etc. Poplar planting also helps in protecting agriculture crops from scorching sun, desiccating effect of wind and also control soil erosion.

**Fig. 6.1:** Poplar based agroforestry system (GBPUAT, Pantnagar, Uttarakhand)

## Economics of Poplar based agroforestry system

The major objective of agroforestry is to optimize production and economic returns per unit area, while following the principle of sustainable development. The economic evaluation of poplar plantations grown on govt. land by forest department and by cultivators on their own farms was conducted at Forest Research Institute, Dehradun (Mathur and Sharma, 1983). The annual returns were discounted at 12, 15, 20, and 25 percent. The NPV (net present value) per hectare of poplar plantation with and without agricultural crops in the forest land and with agricultural crops on farm lands at 8, 10, and 12 years rotation was calculated. Both NPV and benefit cost ratio (B:C) were found maximum at 12 percent discount rate with 8 years rotation. It was concluded that cultivation of agricultural crops with poplar at 8 years rotation ensures higher economic returns.

Poplar-wheat-fodder based agroforestry system is a great success in Punjab. This system is economically viable and much more profitable in comparison to the prevalent rice-wheat system commonly followed by majority of farmers. The comparative economic study of poplar-based agroforestry model in block and boundary plantation revealed that these were 2.8 and 1.6 times more profitable than rice–wheat rotation, respectively.

The income from poplar plantations raised under agroforestry are estimated to provide a gross return of Rs. 4.11 lakh (Rs. 19,500/ ha from agriculture and Rs. 3,91,500 from trees) or a surplus of Rs. 2.78 lakh per ha in eight years with an allowance for interest on investment @ 15 per cent. The returns being substantially higher than raising pure agriculture several fodder crops and medicinal herbs besides other vegetables and some cereal crops. Growing of poplars in orchards has gained popularity in recent years.

**Table 6.3:** B:C ratio analysis of different Poplar based agroforestry system in India

| Agro-forestry system | B/C ratio | References |
|---|---|---|
| Poplar–Sugarcane–Turmeric | 3.06 | Kumar *et al.*, 2004 |
| Poplar–Sugarcane–Wheat–chari | 3.48 | |
| Poplar–Sugarcane–Wheat–chari–Potato–maize–Bajra | 2.58 | |
| Poplar–Sugarcane–Potato–Barseem–chari | 3.02 | |
| Poplar–Paddy–wheat | 2.42 | |
| Poplar–Sugarcane–Wheat–paddy | 2.74 | |
| Poplar +Sugarcane (6th year) | 1.96 | Chaudhary, N. P and Chaudhary, G., 2012 |
| Poplar +Sugarcane (5th year) | 1.76 | |
| Poplar + wheat (4th year) | 1.57 | |
| Poplar + wheat (3rd year) | 1.40 | |
| Poplar + wheat (2nd year) | 1.25 | |
| Poplar + wheat (1st year) | 1.12 | |
| Poplar + wheat(zero year) | 1.00 | |
| Poplar +wheat (5th year) | 1.30 | Rani *et al.*, 2016 |
| Poplar + wheat (4th year) | 1.11 | |
| Poplar + wheat (3rd year) | 0.76 | |
| Poplar + wheat (2nd year) | 0.93 | |
| Poplar + wheat (1st year) | 1.04 | |
| Sole fennel | 1.24 | Rathee *et al.*, 2017 |
| Poplar + fennel (5 × 4 m) | 1.27 | |
| Poplar + fennel (10 × 2 m) | 1.37 | |
| Poplar + fennel (18 × 2 × 2 m) | 1.66 | |

| Agro-forestry system | B/C ratio | References |
|---|---|---|
| Sole ajwan | 0.70 | |
| Poplar + ajwan (5 × 4 m) | 1.10 | |
| Poplar + ajwan (10 × 2 m) | 1.20 | |
| Poplar + ajwan (18 × 2 × 2 m) | 1.32 | |
| Poplar + Sorghum–berseem (5 × 4 m) | 2.20 | Chavan, S. B. and Dhillon, R. S. 2019 |
| Poplar + Cowpea–wheat (5 × 4 m) | 1.92 | |
| Poplar + Sorghum–berseem (10 × 2 m) | 2.22 | |
| Poplar + Cowpea–wheat (10 × 2 m) | 1.98 | |
| Poplar + Sorghum–berseem (18 × 2 × 2 m) | 1.93 | |
| Poplar + Cowpea–wheat (18 × 2 × 2 m) | 1.70 | |
| Sole Sorghum-berseem | 1.13 | |
| Sole Cowpea-wheat | 0.9 | |
| Poplar+Maize (baby corn Variety HIM 123) in open field | 1.10 | Singh *et al.*, 2019 |
| Maize (baby corn Variety DHM 107) in open field | 0.99 | |
| Poplar + Maize Once per week (leaf litter removal) | 0.66 | |
| Poplar + Maize Once per two week (leaf litter removal) | 1.47 | |
| Poplar + Maize Once per three week (leaf litter removal) | 1.95 | |
| Poplar + Maize No removal | 3.07 | |
| Poplar + Maize Twice per week (leaf litter removal) | 0.20 | |
| Poplar + Maize Twice per two week (leaf litter removal) | 0.70 | |
| Poplar + Maize Twice per three week (leaf litter removal) | 1.20 | |
| Poplar + Maize No removal. | 2.90 | |

## Conclusion

Owing to its easy propagation/easy availability of quality planting material, casting less shade on agricultural crop due to small canopy, fast growing rate, high productivity (up to 50 cubic metre wood/ha/year) in 6–12 years rotation, good market opportunity, farmers are going for large scale boundary plantation of poplar. Keeping in view the ever-increasing demand of poplar wood and the interest of farmers in poplar cultivation, selected promising clones have been developed.

## References

Bansal AK, Prasad TN and Mathews KC 1999 Plywood from plantation species-*Populus deltoides* (poplar). Paper presented at the National Seminar on Poplar at FRI, Dehradun. 25–27 November 1999.

Chaudhary NP and Chaudhary G 2012 Poplar Culture on Farmland: Farmer's Experience from Uttar Pradesh. *Forestry Bulletin,* **12**(1) 68–74.

Chauhan SK, Brar MS and Sharma R 2012 Performance of poplar (*Populus deltoids* bartr.) and its effect on wheat yield under agroforestry system in irrigated agro-ecosystem, India. *Caspian J. Environ. Science* **10**(1) 53–60.

Chavan SB and Dhillon RS 2019 Doubling farmers' income through *Populus deltoids* based agroforestry systems in northwestern India: an economic analysis. *Current science*, **117**(2) 219–226.

Kumar R, Gupta PK and Gulati A 2004 Viable agroforestry models and their economics in Yamunangar district of Haryana and Haridwar district of Uttaranchal. *Indian Forester* **130**(2) 131–148.

Rani S, Rajasekharan A, Benbi DK and Chauhan SK 2016 Economic Evaluation of Different Land Use Systems in North Western Region of Punjab, India. *Forest Res.* **6**(1) 192. doi:10.4172/2168-9776. 1000192.

Rathee P, Kaushik N, Khajuria S, Singh P and Manjeet 2017 Performance of Fennel and Ajwan as Intercrops under Different Spacings of Poplar Plantations in North–Western, India. *Int. J. Pure App. Biosci.* **5**(1) 901–907.

Sharma KK 1996 Agroforestry in farming systems development. *Indian Forester* **122**(7) 547– 559.

Singh AP, Bhandari RS and Verma TD 2004 Important insect pests of poplars in agroforestry and strategies for their management in northwestern India. *Agroforestry Systems* **63** 15–26.

Singh B and Sharma KN 2007 Tree growth and nutrient status of soil in a poplar (Populus *deltoids* Bartr.) based agroforestry system in Punjab, India. *Agroforestry Systems* **70** 125–134.

Singh P and Singh S 1986 Insect–pests and diseases of poplars. Forest Research Institute, Dehradun, 74 page.

Singh V, Daniel S, Agarwal YK and Kumar B 2019 Performance Analysis and Economics of Leaf Litter Fall and Baby Corn Intercrop Under Poplar Based Agroforestry System. *Int. Arch. App. Sci. Technol.* **10**(4) 56–60.

Verma RK 1993 Studies on growth and performance of *Populus deltoides* Bartr. in Punjab. M.Sc. thesis. Punjab Agricultural University, Ludhiana, 154 page.

## CHAPTER - 7

# Eucalyptus based Agroforestry System

### EUCALYPTUS

*Eucalyptus* is the third most planted woody species in the world after Pinus and Cunninghamia. Eucalyptus belongs to family of Myrtaceae with 300 species of genus, mostly found in tropical region is a native to Australia. Majority of species and sub species are naturally found in Australian continent and closely adjacent islands. About three quarter of Australian forests are eucalypt forests. The name eucalyptus is derived from the Greek words "eu" meaning very, well, good, true, or beautiful and "kalypto" meaning hidden, concealed, or covered referring to the flower buds of the species which are covered with a cup-like membrane structure with a hard lid, called operculum, that is thrown off as the flower expands. *Eucalyptus spp.* grows under a wide range of climatic and edaphic conditions in their natural habitats. It is a fast growing tree in world and attains great height (one of its species attained height of 480 feet). It is planted for the purpose of fuel wood, timber, making poles, for biomass and has also got medicinal uses. Essential oil of eucalyptus is called "nilgiri ka tel" and is extensively used in Ayurveda. Flowers of eucalyptus tree are good source of nectar for honey bees. Other names for eucalyptus are nilgiri tree, gum tree, or safeda.

### Distribution

Maximum area under *eucalyptus* spp. plantation is in Brazil (41.17% of global), followed by China (32.61%), Chile (8.87%), Sudan (6.54%), Australia (6.47%), India (2.86%), Argentina (1.34%) and the least area under eucalyptus is in Myanmar (0.09%). In India Eucalyptus is the second most widely planted species after Teak. It has been planted all over India on a large scale, except north-eastern states, in Andhra Pradesh, Tamil Nadu, Punjab, Uttar Pradesh, Haryana, Bihar, Gujarat, Maharashtra, Kerala, Karnataka and Madhya Pradesh.

### Botanical Characters

Eucalyptus is a fast-growing, medium to tall tree growing up to 20–50 m height and attains as much as 2 m diameter. It has a deep tap root system with mycorrhizal associations which increases its nutrients and water uptake abilities. Eucalyptus tree has a smooth silvery white trunk while stems of the seedlings and coppice shoots are quadrangular. The leaves are leathery in texture, hang obliquely or vertically on peduncle and have glands containing aromatic oil. Flower buds are covered with a conical cup– like membrane, which is thrown off when the flower expands. Flowers are arranged in cymose panicles. The fruit is a capsule and is surrounded by a woody, cup-shaped receptacle and contain numerous minute seeds.

### Varieties

New species like *Eucalyptus camaldulensis, Eucalyptus globules, Eucalyptus citriodora*, and Eucalyptus clones such as FRI 4, FRI 6, IFGTB–EC1, -EC2, -EC3 and -EC4.

### Climate requirement

Eucalyptus is grown over a wide range of latitude between 60 °N to 60 °S and altitude ranging from sea level to 3000 m. It can be easily grown on infertile soils and is tolerant of extreme drought and high temperature. It is suitable tree species for the areas receiving rainfall as low as 250 mm to area receiving high rainfall up to 1250 mm.

### Soil

Eucalyptus species occurs on a variety of soil types from red or black soils to sandy alluvial soils. It can also grow well in salt affected areas. Suitable soil pH range lies in acidic (6.5 to 2.0).

### Propagation Technology

Eucalyptus is mainly propagated by seeds, cuttings and can also be propagated through grafting and micropropagation.

**Through Seeds**– Seed should be sown under shade on a well–drained and sterilized bed and should be covered sparingly with inert material (*e.g.* sand). Seeds should neither be sown too deep, nor too shallow to prevent exposure while watering. The seeds take 4 to 5 days to germinate. After the seedlings attain height of 5 to 7 cm (3–5 months after sowing), they are lifted from the mother bed and transplanted in poly bags. Partial shade is needed up to six weeks from transplanting. In four to five months from transplanting, plants reach plantable size of 30 cm. Seedlings require periodic watering in the early stages of development. About 1–2 lakhs seedlings can be easily obtained from 1 kg of seed in nurseries as 100–500 germinants can come per gram of seeds.

**Through Cuttings**– Choose 4 inch long mature shoots during June/July and dip the bottom tips of the cuttings in rooting hormone (4000ppm IBA + 4000ppm NAA) for about 30 seconds. Eucalyptus cuttings should have at least one budding leaf but sprouting leaves should be broken off. A pot is filled with perlite and the cuttings are put down into the medium with the end dipped in rooting hormone covered. A saucer kept beneath the pot is filled with water and the pot is allowed to absorb water until moistened through its bottom hole. Then the pot is covered with a plastic bag and placed in a warm location (80–90 °F or 27–32 °C). Eucalyptus cuttings should be kept moist and in around four weeks the roots will come and the cuttings will be ready for transplanting.

### Time of sowing

Planting should be done from June to October.

### Spacing

Spacing of 1. 5m × 1. 5m with about 4444 plants/ha or 2m × 2m with 2500 plants/ha is selected for high density planting. 2m × 2m is the most widely used spacing. Intercrops can be taken in the early years of plantation. However, wider spacing of 4m × 2m or 8m x 1m (about 1250 plants/ha) and 6m × 1.5m (1111 plants/ha) can be opted when agroforestry is to be practiced. Shade loving crops like turmeric and ginger or medicinal plants can be taken as intercrops.

### Planting

Seedlings are planted in pits with the onset of monsoon. For planting of eucalyptus require well drained, deep, fertile soils and sandy soils also may be good if provided facility of sufficient irrigation is available. Choice of proper species for a particular site is also an important aspect for the success of eucalyptus plantation. In manual planting, pits of 15 × 15 × 15 or 45 × 45 × 45 or 60 × 60 × 60 $cm^3$ can be dug depending on the locality, soil and irrigation facilities. In waterlogged areas, planting on raised mounds up to 1m high is done in certain areas of Bihar, West Bangal, Punjab and Uttar Pradesh.

### Irrigation

Irrigation should be applied immediately after transplanting of seedlings in the main field. No irrigation is required if there is enough rain but life–saving irrigation is needed in case of monsoon failure or dry spells in initial stage. Though eucalyptus is a drought tolerant tree, nearly 25 irrigations is required in entire growth period for optimum yield. Irrigation is needed mostly in summer season and to some extent in winter season.

### Nutrient management

At the time of planting, apply neem-based nutrients like neem cake, neem coated urea etc. along with Phosphate @ 50 gm and vermicompost @ 250 gm/pit. Neem based nutrients protect roots of seedlings from termite damage. During first and second year, apply 50gm of NPK (17:17:17) fertilizer per plant. Also take hand weeding operation and keep check on weed growth to reduce nutrient loss by weeds.

**Table 7.1:** Plant Protection of Eucalyptus

| Pest | Control |
|---|---|
| **Termite:** In young plant termite is most dangerous insect and it damage crop to great extent. | To protect crop from termite attack apply Nimbicides @ 2 ml/L of water. |
| **Gall:** It causes drying of leaves, growth retardation, poor stem formation. New leaves appeared in gall infected plants are stunted. | If infection observed, remove and destroyed gall affected plants. Use gall resistant varieties. |
| **Stem Canker:** If infestation is observed | Apply Bordeaux mixture in root portion. |

## Yield

Generally, the rotation age for pulpwood production is 4 years and for plywood 6–8 years. The expected yield is up to 60–70 t/ha in 6 year rotations. The mean annual increment of the species is 20–60 cu m/ha/year. From clonal plantation yield of 50 MT to 76 MT in five year rotation is obtained whereas from origin plantation 30 MT to 50 MT is obtained. Yield greatly depends upon management practices, plant density, climate etc.

## Utilization

*E. tereticornis* has strong, hard and durable heartwood, with about 1100 kg per cubic m density. Wood is used to extract veeners in the preparation of plywood, paper and pulp industry. The calorific valuc of a 4 year eucalyptus is about 1920 MJ/kg which used in fuel wood and produce good charcoal. Green saw wood is used for boxes, pallets, bins, plank mouldings and scaffolds; the dried ones are used as raw material for manufacture, moldings, and carpentry. Its poles are good for transmission purposes and are also used in construction of dwelling houses, work sheds and in mines. The leaves are used in the production of eucalyptus oil and can be promoted as a bio-pesticide. The wood and bark of the tree have a tannin content of 6–12% and 3–15%, respectively. Eucalyptus species are rich in nectar and pollen for apiculture. It is also used as ornamental trees suitable for parks and avenue plantations.

## Eucalyptus in Agroforestry Systems

At a spacing of 3 m × 2 m, intercropping can be carried out for one year. If irrigation is available then shade loving crops can be cultivated during the second year also. For block plantation, the distance between the spacing should be 3 m × 1.5 m, which will hold 2222 plants per hectare and allow rising of intercrops in the first 2 years in a cycle of 4 years rotation. In agri-silviculture system a wide range of crops can be grown up to 3 years when the spacing is kept at 5 m × 2 m. Intercrops such as cotton, tobacco, chillies, black gram, green gram are widely grown in eucalyptus plantation during the initial years. An experiment was initiated at ICAR–CAFRI, Jhansi, in 2003 to assess the production potential of intercrops (black gram and wheat) with various clones of eucalyptus planted at two spacings,

5 m × 4 m and 10 m × 2 m under boundary plantation. The grain yield of black gram was 1.0 to 1.25 t/ha during the first and second year, but it dropped to 0.16 to 0.18 t/ha after the sixth year. In case of wheat, the grain yield was 2.17 to 2.56 t/ha during the initial years, but it also dropped to 0.78 to 0.83 t/ha after the sixth year. In Northern India, exotic species such as eucalyptus and poplar are being raised on an enormous scale as agroforestry trees by farmers on field bunds, within agricultural fields, wasteland etc. The popularity of these species is mainly because they are fast growing, tolerant to pruning and have small canopy which casts minimum shade effects. Eucalyptus plantation started mushrooming in Northern part of India with farmers taking various crops like sugarcane, wheat, maize etc. as an intercrop.

## Case study– Eucalyptus based Agro-forestry models

In India, eucalyptus has been grown on farmlands since sixties, as wind break, as avenue tree on farmers homesteads, cattle shed and on the side of farm approach road. Raising of eucalyptus on field boundaries under different patterns of planting,

Fig. 7.1: Eucalyptus based agroforestry system (CCSHAU, Hisar)

is a common practice adopted by farmers. Eucalyptus, is a very good companion tree for agricultural crops under agroforestry system. Due to its self-pruning ability, eucalyptus grows straight having no branches and hence casts no shade to the associated crop. It's leaves are also easily degradable and enriches soil. However, cases of adverse effects on companion crops are also reported. Apart from some allelopathic effects observed, eucalyptus tree also absorbs large amount of water in a short period of time. But its fast growth habit and huge biomass production resulting in economic returns compensates for these adverse effects. The economic feasibility of Eucalyptus hybrids, both with and without agricultural crops, under varying spacing *viz.* 2.5 m × 2.5 m, 3 m × 1.5 m, 4 m × 2 m and 6 m × 1 m was worked out. The result of the study indicates that 6m × 1m spacing of eucalyptus in a rotation of 8 years with agricultural crops on farm lands gave maximum NPV and B:C ratio of 2.28.

## Economics of Eucalyptus based agroforestry system

Eucalyptus based agroforestry is highly profitable for farmers as its wood is in great demand in India as well as abroad. Eucalyptus based agroforestry system gave higher gross income, net return, B:C ratio than open farming in a number of experiments. Eucalyptus clones yield around 300–400 kg biomass per tree after 5 years. In agroforestry system, total plants per hectare comes around 1111 to 1250. On completion of 5th year, total yield per hectare comes around 400 metric tonnes and with the present market rate of Rs. 6000–7000/ton, it fetches around 20 to 30 lakhs Rs. per hectare.

**Table 7.2:** B:C ratio analysis of different Eucalyptus based agroforestry system in India

| Agro-forestry system | B/C ratio | References |
|---|---|---|
| Eucalyptus (bund planting 2 m) + crop | 2.32 | Mathur *et al.*,1984 |
| Eucalyptus (2.5 × 2.5 m) + crop | 2.08 | |
| Eucalyptus (3 × 1.5 m) + crop | 1.57 | |
| Eucalyptus (4 × 2 m) + crop | 1.88 | |
| Eucalyptus (6 × 1 m) + crop | 2.17 | |
| Eucalyptus (10 × 1.5 m) + crop | 2.20 | Prasad *et al.*, 2010 |
| Eucalyptus (11 × 1 m) + crop | 2.17 | |
| Eucalyptus (7 × 1.5 m) + crop | 2.17 | |
| Eucalyptus (5 × 1 m) + crop | 2.17 | |
| Eucalyptus (3 × 2 m) + crop | 2.07 | |
| Sole Eucalyptus | 2.29 | |
| Sole arable cropping | 1.93 | |
| Eucalyptus + Mango + Arhar | 1.18 | Annual Report |
| Eucalyptus + Mango + Cowpea – Toria | 1.52 | (2017–2018), BCKV, |
| Eucalyptus+ Mango + Greengram – Toria | 1.47 | West Bengal |
| Sole Arhar | 0.99 | |

| Agro-forestry system | B/C ratio | References |
|---|---|---|
| Sole Cowpea – Toria | 1.44 | |
| Sole Greengram – Toria | 1.35 | |
| Eucalyptus + Mango + Pigeon pea | 3.80 | Dhara, *et al.*, 2018 |
| Eucalyptus + Mango + Black gram | 3.40 | |
| Eucalyptus + Mango + Lady's finger | 3.80 | |
| Eucalyptus + Mango + Bottle gourd | 3.90 | |
| Eucalyptus + Mango + Maize | – | |
| Eucalyptus + Mango | 2.50 | |
| Sole Eucalyptus | 2.00 | |
| Sole Mango | 2.20 | |
| Eucalyptus (3 × 3 m) + crop | 1.40 | Dhillon, *et al.*, 2018 |
| Eucalyptus (6 × 1.5 m) + crop | 1.43 | |
| Eucalyptus (17 × 1 × 1 m) + crop | 1.57 | |
| Sole Eucalyptus (3 × 3 m) | 1.12 | |
| Sole crop | 1.22 | |

## Conclusion

Eucalyptus, being a fast growing species, is a very good option for industrial agroforestry. Eucalyptus is one of the most popular choice to be planted along the edges or bunds of farmlands. Silvicultural properties including straightness, narrow crown, self pruning nature, coppicing (a tendency not to spread as a weed), and wide utility of its wood make it popular among the farmers. Wider spacing of eucalyptus is recommended under agroforestry system which ensures high economic returns.

## Reference

Annual Report 2017–18 Yield and economics of *Eucalyptus tereticornis* + Mango based agroforestry system. Regional Research Station (Red & Laterite Zone), Bidhan Chandra Krishi Viswavidyalaya, Jhargram, West Bengal.

Dhara PK, Panda S and Sarkar S 2018 Eucalyptus + Mango fruit based agroforestry system. Paper presented in annual group meeting of AICRP– Agroforestry, 28–30 July 2018, Birsa Agricultural University, Ranchi, Jharkhand.

Dhillon RS, Chavan SB, Bangarwa KS, Bharadwaj KK, Kumar S and Sirohi C 2018 Eucalyptus–based Agroforestry System under Semi–Arid Condition in North–Western India: An economic Analysis. *Indian Journal of Ecology* **45**(3) 470–474.

Mathur RS, Sharma KK and Ansari MY 1984 Economics of Eucalyptus plantations under agroforestry. *Indian Forester* **110**(2) 171–201.

Prasad JVNS, Korwar GR, Rao KV, Mandal UK, Rao CAR, Rao GR, Ramakrishna YS, Venkateswarlu B, Rao SN, Kulkarni HD and Rao MR 2010 Tree row spacing affected agronomic and economic performance of Eucalyptus-based agroforestry in Andhra Pradesh, Southern India. *Agroforest System* **78** 253–267. DOI 10.1007/s10457-009-9275-1

Rajan BKC 1987 Versatile Eucalyptus. Diana Publishers, Bangalore, India, 164–172 pages.

CHAPTER - 8

# Malabar Neem based Agroforestry System

## MALABAR NEEM

*Melia dubia* (Malbar Neem/ Maha Neem/ Bakain) is a fast growing industrial wood species which belongs to the family Meliaceae. It is native to India, South East Asia and Australia where it has been cultivated as a source of firewood. It is a fast growing tree under irrigated condition and produces greater bio-mass in relatively shorter period, can grow up to 8–9 m within two years from planting. This tree has incredible property of self pruning and is termite resistant, farmer friendly, frost and drought resistant and can be mechanically pruned and harvested. It grows naturally in certain parts of the Western Ghats of South India and is found in moist deciduous, evergreen and semi-evergreen forests. It is mostly found growing in the country sides on wastelands. It is an economically important multipurpose tree species predominantly growing in southern India whose cultivation has now expanded to whole country (Parthiban *et al.*, 2009). *Melia dubia* serves as raw material for plywood, pulp and paper industries and also used for making packing case, match splints etc. For plywood production or in paper industry, it can be harvested early *i.e.* within four years of planting whereas with advancing years of maturity (beyond 7 to 15 years) ensures good quality timber that can be used for construction.

### Distribution

*Melia* is indigenous to the Western Ghats of Southern India and is commonly found in moist deciduous forests of Kerala. It also occurs in Himalayas in Sikkim, North Bengal, upper Assam, and Khasi hills, Eastern Ghats, Deccan Plateau and Western Ghats, at the altitudes of 1500–1800 m. It grows rapidly and is used for afforestation. Outside India, it is found in Sri Lanka, Malaysia, Java, China and Australia.

### Botanical description

The tree is deciduous to semi-evergreen, grows up to 25 m height with wide spreading branches of handsome foliage. Bark is dark brown in color, fibrous and peels off in long strips. Leaves are compound, bipinnate to tripinnate, alternate, pulvinate; rachis terete to 30 cm long; petiole 0.3–1.2 cm; pinnae 3–8 pairs; leaflets 2–11 on each pinnate, opposite, 4.5–9 × 2–4 cm, ovate elliptic, apex acuminate, base cuneate or attenuate, asymmetric, margin crenate, coriaceous, glabrous when mature; midrib raised above; secondary nerves of 10 pairs, gradually curved; tertiary nerves broadly reticulate. Inflorescence is an 12–20 cm long auxiliary panicle; flowers are small, greenish white in color having sweet smell, and appear in bunches with new flush of leaves during January to March. Fruit is a drupe, ovoid or ellipsoid in shape with longitudinal ridges, becomes pulpy and yellowish on ripening with a sweet smell. Each fruit consists of 3–4 seeds. Fruits have a hard endocarp. Germination is rather difficult as the tough seed coat does not allow water to penetrate easily. Around 250–320 seeds weigh one kilogram (Nuthan *et al.* 2009). Its leaves fall from February to May, flowers January to March and fruits from November to February. Leaves are rich in essential oils,

## Climate requirement

Absolute maximum and minimum shade temperature in its natural habitat varies from 37.5–47.5 °C and 0–15 °C, respectively. *Melia dubia* grows well in moist regions, with a mean annual rainfall exceeding 800–1000 mm. The mean relative humidity in July varies from 70–90% and in January from 50–80%.

## Soil

The tree grows on and can be cultivated on a variety of soils; however, it grows well in deep, fertile sandy loam, red and lateritic soils with an annual rainfall of 800 mm and above. It requires a low supply of water, initially on a weekly basis.

## Variety

- Malaivembu MTP 1 (plywood)
- Malaivembu MTP 2 (pulp and plywood)

## Propagation technique

**Seed sowing:** Seeds are best to be sown in March–April. The seed coat of *Melia dubia* is very tough which doesn't allow easy water penetration and hence germination. The seeds should be partially rotten or decomposed for loosening the seed coat by placing the them in heap of cow dung. The softness of the seeds is tested periodically to find out the exact time for peeling the external coat of the seeds. It takes about two months. When the outer seed coat becomes soft, the coat can be removed and by making shallow longitudinal cuts on the inner hard layer with the help of a sharp knife, seeds are taken out. This should be done very carefully, so that there isn't any slight damage to kernels. Two to three seeds can be extracted from each kernel. The seeds look similar to sunflower seeds. After cleaning and drying, seeds should be sown in open nursery on raised beds, 5cm apart from each other.

For better germination, a medium of soil and farm yard manure mixed in 1:1 or 2:1 ratio. About 6–7 kg of dried drupes (around 1500 in number) are required for one standard nursery bed of 12 m × 1.2 m size. After sowing, the beds need to be watered regularly, twice a day. At places where daytime temperature is not very high, or where nursery beds are in shade, the beds should be covered with a tarpaulin sheet to maintain temperature. Germination occurs within 90 days.

At two leaf stage, seedlings are uprooted carefully without damaging the root–system and planted into polythene bags of 15 cm × 8 cm dimension. The bags are made out of thermoplastic material and are filled with a mix of red soil, fine sand and organic manure. The seedlings will grow up to 60 cm height (2 ft.) in another 30 days and are ready to be planted in the main field.

### Vegetative propagation

Juvenile stem cuttings and coppice shoots respond well to 1000–2000 ppm IBA (liquid formulation). Coppice from older trees roots better. Cuttings of pencil thickness need to be taken for propagation as thin shoots are susceptible to root rot. The cuttings can be placed on sand medium and watered twice daily. There must be proper drainage as water logging destroys the cuttings. Weather also plays a major role in the rooting of cuttings. Drier seasons are conducive for rooting and about 75 per cent rooting can be obtained in dry season.

### Transplanting

The rooted saplings are planted with the onset of the monsoon or during monsoon. However, it can be successfully grown in regions having very less or no rains with supplemental irrigation. Pits of 60 cm × 60 cm × 60 cm size are dug mechanically or manually by using a hand–hoe. A spacing of 3 m × 3 m is generally recommended for sole planting. A wider spacing of 5 m × 5 m or 8 m × 8 m is followed when agroforestry is to be practiced. Pits are filled with soil, fine sand and well decomposed manure. Seedlings are placed upright in the centre and soil is pressed well around the seedling base.

### Irrigation

For fast growth of the trees regular irrigation is needed. Growth is slow under rainfed conditions. In initial stage daily watering hastens growth. The tree responds well to irrigation at once in every 10–15 days during non rainy season.

### Fertilizer Requirements

Growth is enhanced with the application of fertilizers once in three months for the first three years. Application of 25–50 g N, P, K mixture per tree should be applied twice a year enhance the growth. The fertilizer requirements can be scheduled on need basis depending on the growth and development of the tree.

### Pruning

Initially, periodical pruning at 15 days intervals is done when the plants attain 2 m height. This results in straight trunk up to 10 m height. The side branches are pruned and the nodal buddings grown–in between the main trunk and the side branches are nipped. Pruning every six months controls branching. The bole is straight, round, without any knots and without any buttress.

## Harvesting

After 7 to 12 years the trees offer good timber, which can be cut and sold easily and profitably. For quick return, trees can also be sold at three years of age to paper industries. Ratooning is a method of harvesting a crop which leaves the roots and the lower parts of the plant to give the ratoon crop. Trees are cut just above the ground level and many side suckers emerge from the basal region of the stem. Out of these only one sucker is allowed to grow as ratoon crop. The main benefit of ratooning is that it decreases the cost of field preparation. Plant grown from ratoon also matures earlier in the season as ratoons grow fast and attain sufficient thickness within short time compared to freshly planted seedlings. However, this method cannot be used endlessly as the yield of the ratoon crop decreases after each cycle. Ratooning is most often used in trees which give steady yield for three years under favorable conditions. Three successive ratoons can be allowed for optimum production with maximum returns without affecting the productivity or profitability of tree farming.

## Yield

The tree attains a volume of 0.15 to 0.20 cu m in 5 to 6 years and 0.42 cu m after 15 years. The wood fetches a price of Rs 7300/– per ton for billets of girth 50–120 cm Rs 12,000 per cubic m from $5^{th}$ year onwards for trees which have attained a girth >120 cm (Warrier, 2014). Growth rate ranges from 2 to 2.5 m per year in intensively managed plantations to as low as 0.6 to 0.8 m per year in unmanaged plantations.

## Utilization

It is a good source of secondary timber and one of the most preferred species for plywood industry. The wood is also used for making packing cases, ceiling planks, building purposes, agricultural implements, pencils, match boxes, splints, catamarans, musical instruments and tea boxes as its wood is anti–termite. Thus, the species has a ready and assured market due to its multipurpose utilities and is in high demand by the plywood industries.

## Malabar Neem in Agroforestry System

*Melia dubia* is highly suitable under agroforestry system and is a promising tree species for generating higher income in the semi-arid climates. A variety of crops groundnut, chilli, turmeric, green gram, blackgram, papaya, banana, melon, foxtail millet and sugarcane can be successfully grown in combination with *Melia dubia* as intercrops. The species performs exceedingly well when planted on bunds, attaining the harvestable size within four years.

*Melia dubia* can also be grown with other trees such as Teak, Nutmeg and Coconut to fill the inter space in a Plantation to utilize the empty space advantageously.

**Fig. 8.1:** Malabar Neem based agroforestry system (TNAU, Coimbtore)

## Economics of Malabar Neem based agroforestry system

Jilariya *et al.*, 2019. reported that sole planting of *M. dubia* (4 m × 2 m) gave maximum BCR (3.71) among all land-use systems considering *Aloe vera* fresh leaf production in to account. The value addition due to pulp production provided higher net returns and BCR under all the systems. The net present value (NPV), internal rate of returns (IRR@10%), payback period (PBP) and BCR analysis expressed that among *M. dubia–A. vera* silvi-medicinal and sole cropping systems (either *A. vera* or *M. dubia*), the most economically viable system could be *M. dubia* (2 m × 2 m) + *A. vera* system. The economic analysis underpinned that NPV, IRR, PBP and BCR values are lesser on raw product (fresh *A. vera* leaf) production compared to pulp production.

Ashalatha, *et al.*, 2015. also reported maximum gross income (Rs. 87,000 $ha^{-1}$), net return (Rs. 67,000 $ha^{-1}$) and benefit cost ratio (4.30) in *M. dubia* + brinjal among various other combinations and sole cropping. The performance of brinjal under *M. dubia* might be due to higher productivity and better return when compared to pure crops. *M. dubia* plantations also contributed towards higher carbon sequestration.

**Table 8.1:** Net income of Malabar Neem after 8 years

| Cost of *Melia dubia* (per acre) | Rupees (Rs.) |
|---|---|
| Preliminaries / Pitting / Farmyard manure / Planting etc. | Rs. 20,000 |
| Cost of 400 rooted plants: Rs. 200/– each. | Rs. 80,000 |
| Cost of Drip line | Rs. 40,000 |
| Total Expenditure for 1year | Rs.1,40,000 |
| ADD: Weeding / Watering / Manure etc. for years | Rs. 40,000 |
| **Total Cost** | **Rs.1,80,000** |
| **Income: (After 8 years)** | |
| Average yield 20 Cubic feet wood per Tree × 400 Trees = | 8000 Cft. |
| Realisation @ Rs.200/ cft. × 8000 = | Rs.16,00,000 |
| **Net Income: (16,00,000–1,80,000) =** | **Rs.14,20,000** |

**Table 8.2:** B:C ratio analysis of different Malabar Neem based agroforestry system in India

| Agro-forestry system | B/C ratio | References |
|---|---|---|
| *Simarouba glauca* + crop | 2.86 | Anusha, S., 2012. |
| *Melia dubia* + crop | 3.96 | |
| *Azadirechta indica* + crop | 2.34 | |
| *Melia azedarach* + crop | 2.64 | |
| *Pongamia ponnata* + crop | 2.38 | |
| *Madhuca latifolia* + crop | 2.50 | |
| *Callophyllum inophyllum* + crop | 2.40 | |
| Sole crop | 2.15 | |
| *M. dubia* + greengram | 3.10 | Ashalathal *et al.*, 2015 |
| *M. dubia* + blackgram | 3.16 | |

| Agro-forestry system | B/C ratio | References |
|---|---|---|
| *M. dubia* + cowpea | 2.46 | |
| *M. dubia* + bhendi | 2.53 | |
| *M. dubia* + small onion | 2.10 | |
| *M. dubia* + growndnut | 2.62 | |
| *M. dubia* + sesame | 2.80 | |
| *M. dubia* + brinjal | 4.30 | |
| *M. dubia* + chillies | 2.50 | |
| *Melia dubia* as hedge in cotton | 2.05 | Kulkarni, S. G., 2017 |
| *Melia dubia* as a border crop cotton | 2.61 | |
| *Melia dubia* + Papaya (1:1) | 1.25 | |
| *Melia dubia* as a border crop papaya | 1.53 | |
| *Melia dubia* + Pomegranate (1:1) | 3.39 | |
| *Melia dubia* as a border crop Pomegranate | 3.21 | |
| *Melia dubia* + Sweet lime (1:1) | 2.31 | |
| *Melia dubia* as a border crop Sweet lime | 2.58 | |
| Sole *Melia dubia* | – | |
| Sole Cotton | 3.02 | |
| Sole Papaya | 2.15 | |
| Sole Pomegranate | 3.13 | |
| Pomegranate + Papaya(1:1)– Pomegranate equivalent yield | 2.64 | |
| Sole *Cymbopogon* spp. | 2.97 | Mohanty *et al.*, 2017 |
| *M. dubia* (2 × 2 m) + *Cymbopogon* spp. | 3.86 | |
| *M. dubia* (3 × 3 m) + *Cymbopogon* spp. | 4.41 | |
| *M. dubia* (3 × 2 m) + *Cymbopogon* spp. | 4.33 | |
| *M. dubia* (4 × 4 m) + *Cymbopogon* spp. | 3.14 | |
| *M. dubia* (4 × 2 m) + *Cymbopogon* spp. | 3.95 | |
| Sole *M. dubia* (2 × 2 m) | 0.25 | |
| Sole *M. dubia* (3 × 3 m) | 5.44 | |
| Sole *M. dubia* (3 × 2 m) | 1.59 | |
| Sole *M. dubia* (4 × 4 m) | 4.02 | |
| Sole *M. dubia* (4 × 2 m) | 1.64 | |
| *M. dubia* + greengram | 2.10 | Shudhagar *et al.*, 2017 |
| *M. dubia* + blackgram | 2.80 | |
| *M. dubia* + cowpea | 2.80 | |
| *M. dubia* + bhendi | 2.10 | |
| *M. dubia* + small onion | 2.00 | |
| *M. dubia* + growndnut | 2.30 | |
| *M. dubia* + gingelly | 1.60 | |
| *M. dubia* + brinjal | 3.10 | |
| *M. dubia* + chillies | 2.50 | |

| Agro-forestry system | B/C ratio | References |
|---|---|---|
| *M. dubia* (8 × 5 m) + Finger millet | 18.64 | Bhaskar *et al.*, 2019 |
| *M. dubia* (10 × 5 m) + Finger millet | 17.30 | |
| *M. dubia* (12 × 5 m) + Finger millet | 17.85 | |
| *M. dubia* (16 × 5 m) + Finger millet | 15.13 | |
| *M. dubia* (20 × 5 m) + Finger millet | 13.62 | |
| *M. dubia* (24 × 5 m) + Finger millet | 12.06 | |
| Sole Finger millet | 2.14 | |
| *M. dubia* + *A. vera* (2 × 2 m) | 3.82 | Jilariya *et al.*, 2019 |
| *M. dubia* + *A. vera* (3 × 2 m) | 3.26 | |
| *M. dubia* + *A. vera* (4 × 2 m) | 3.25 | |
| Sole *A. vera* | 1.32 | |
| Sole *M. dubia* (2 × 2 m) | 2.66 | |
| Sole *M. dubia* (3 × 2 m) | 3.59 | |
| Sole *M. dubia* (4 × 2 m) | 3.71 | |
| *M. dubia* + Pearl millet (control ) (6 year old) | 5.36 | Chandana *et al.*, 2020 |
| *M. dubia* + Pearl millet 100% RDF through normal urea | 6.20 | |
| *M. dubia* + Pearl millet 100% RDF through neem–coated urea | 7.24 | |
| *M. dubia* + Pearl millet RD N + 25% N through poultry manure | 5.48 | |
| *M. dubia* + Pearl millet 75% RD N + 25% N through FYM | 5.97 | |
| *M. dubia* + Pearl millet 75% RD N + PGLM @ 10 t /ha | 5.79 | |
| *M. dubia* + Pearl millet 75% RD N + Azotobacter@ 500 g/ ha | 5.39 | |
| Sole Pearl millet without trees (80–40–30 NPK kg/ ha) | 2.96 | |

## Conclusion

*Melia dubia*, being one of the fastest growing tree species, is a promising tree under agroforestry and *M. dubia* based agroforestry system is a sustainable option than the sole agricultural cropping, not only for higher production and better income but also for climate change mitigation. *Melia dubia* combined with agricultural crop ensures profitability, productivity and sustainable utility.

## Reference

Anusha S 2012 Performance of fingermillet and carbon sequestration in agroforestry system. M.Sc. thesis, University of Agricultural Sciences, GKVK, Bangalore, 142 page.

Ashalathal A, Divya MP and Ajayghos V 2015 Development of Suitable *Melia dubia* based Agroforestry Models for Higher Productivity. *Madras Agric. J.* **102**(7–9) 264–267.

Bhaskar V, Hanumanthappa DC, Bhavya V, Nagaraju and Somashekar KS 2019 Growth, Yield and Economics of Finger Millet (*Eleusine coracana*) in *Melia dubia* based Agro Forestry System. *Int. J. Curr. Microbiol. App. Sci.* **8**(5) 1945–1950.

Chandana P, Lata AM, Khan MAA and Krishna A 2020 Climate change smart option and doubling farmer's income through *Melia dubia* based agri-silviculture system. *Current Science* **118**(3) 444–447.

Jilariya DJ, Thakur NS, Singh N and Gunaga RP 2019 Economics of cultivation of *Melia dubia* Cav.-*Aloe vera* L. silvi-medicinal model. *Indian J. of Agroforestry* **21**(2) 35-40.

Kulkarni SG 2017 Suitability study on *Melia dubia* based agroforestry system in north Karnataka. *Bull. Env. Pharmacol. Life Sci.* **6**(12) 49–52.

Mohanty S, Thakur NS, Gunaga RP, Dobriyal MJ and Desai BS 2017 Value Addition in *Cymbopogon spp.* to Enhance the Financial Flows from *Cymbopogon spp.-Melia dubia* Cav. based Silvi-medicinal and Sole Cropping Systems. *Indian Journal of Ecology* **44**(6) 812–816.

Parthiban KT, Bharathi AK, Seenivasan R, Kamala K and Rao MG 2009. Integrating *Melia dubia* in Agroforestry farms as an alternate pulpwood species, *APA News* **34** 3–4.

Shudhagar RJ, Divya MP, Ramah K and Sivakumar K 2017 Pulpwood species based agroforestry models. In book Parthiban KT and Seenivasan R plantation and agroforestry-pulpwood valve chain approach, Scientific Publishers (India), 293 page. ISBN-9789386347985

Warrier RR 2014 Cultivation techniques for *Melia dubia*. In: Transfer of Tree Cultivation Technologies to Krishi Vigyan Kendras (KVKs) of Tamil Nadu and Puducherry. Institute of Forest Genetics and Tree Breeding, Coimbatore, India, 16–18 pages.

Chapter - 9

# Sandalwood based Agroforestry System

## SANDALWOOD

*Santalum album* L., commonly known as True sandalwood, East Indian sandalwood or chandan is a highly valuable tree for centuries, belonging to the family Santalaceae. Its wood is one of the most expensive woods in the world owing to its long-retained fragrance. It is indigenous species of South India, although is distributed all over the country but is found to be naturally growing in forests of western ghats and more than 90% area (around 8300 sq. km.) comes under Karnataka and Tamil Nadu. Sandalwood trees in India are owned by government and their cutting is controlled, but many trees are illegally cut down and smuggled due to its high price. Overharvesting coupled with its slow growth has made it an endangered species. The essential oil obtained by steam distillation of the wood has significant place in perfumery industries/market and is used for making perfumes, soaps, candles, incense, and folk medicines. Other species of genus *Santalum* has been found and grown in other countries but the Indian Sandalwood has retained its dominance in the market over others due to its quality. Sandalwood is known as a semi-parasitic plant (hemi-parasitic), associated with host plants through "haustoria" for: nutrient fulfillment (Kharisma and Suriamihardja 1988), and shade at young stage. Nitrogen (N), phosphorus (P) and amino acids are obtained from host plants, while Calcium (Ca) and Potassium (K) are taken from the soil.

### Botanical characters

True Sandalwood (*Santalum album* L.) is a slow growing evergreen tree attaining a height of 4 to 20 meters and a girth of 1 to 2.4 meters. The shape of the leaves vary with six morphological types recognized widely. Leaves are opposite and decussate, sometimes show whorled arrangement. They are Ovate, Lanceolate, Elliptic, Linear, Big and Small. Stem is smooth when young but turns rough with deep vertical cracks in mature trees. The wood is hard, heavy, and fine-grained. Sapwood is white and scentless while yellowish to brown and strongly scented heartwood is developed after 10 years of age. Trees start flowering from 3 to 7 years of age. Flowers are purplish brown, unscented and are borne on axillary or terminal cymose panicles. The floral organs develop in acropetal succession. It flowers and fruits twice a year during March–April and September–October. However, some trees flower only once a year and some do not flower regularly. Fruit is a drupe, single seeded, round to oblong in shape and purplish in colour when fully mature. About 6000 seeds make 1 kg. Seeds are collected directly from the tree. Both tree and roots contain a yellow aromatic oil, called sandalwood oil. Plants are harvested between 10–30 years of age because the potency of essential oils increase with age and can even take 30 years for the heartwood to reach an economically useful thickness.

## Distribution

*Santalum album* or true sandalwood is indigenous to South India and is distributed in the peninsular India, parts of Malaysia, Australia, New Zealand and Polynesia extending to the Hawaiian Archipelago and Juan Fernandez Islands. In India it is distributed in the dry scrub forest of Salem and Coimbatore region of Tamil Nadu, Mysuru and Coorg in Karnataka and Nilgiris up to 900 m. of altitude. It is also found in Andhra Pradesh, Kerela, Maharashtra, Madhya Pradesh, Bihar, and Gujarat.

## Varieties

There are more than 56 species and varieties of 'Santalum' mentioned in the literature, based on the morphological characters in the World *e.g. S. ellipticum, S. freycinetianum, S. paniculatum,* and *S. spicatum.*

## Climate requirement

*Santalum album* is found in a range of climatic regions from coastal dry forests up to 900 m elevation. Normally sandalwood requires humid & hot climate but it can grow well in temperature ranges from 12 °C to 38 °C and annual rainfall between 500 and 3000 mm.

## Soil

It grows best in sandy or stony red soils, but can grow on a wide range of soil with slightly alkaline pH ranging from 6.5 to 8.5.

## Seed technology

*Santalum album* flowers and fruits twice a year during March–April and September–October. However, some trees flower only once a year and some do not flower regularly. Trees can start flowering from 3 years of age. Seed production is generally good in one of the seasons. About 6000 seeds weigh 1 kg. Seeds can be collected directly from the tree.

## Pretreatment

Pretreatment of seeds is needed to loosen the seed coat for easy and better germination. Acid scarification is done by leaving the seeds in concentrated $H_2SO_4$ for 30 minutes and then washing thoroughly in running water. Or the seeds can be soaked overnight in 0. 05% gibberellic

**Host Plants:** Being semi parasitic in nature, sandalwood tree depends on a host plant for absorbing nutrients from soil. *Alternanthera sp.* and crops like *Capsicum frutescens* (cili), *Solanum melongena* (eggplant), *Lycopersicum esculentum* (tomato), *Cajanus cajan* (Pigeon pea), *Vigna radiata* (green gram), *Musa paradisiaca* (banana) can be used as host plant for sandalwood tree in nursery. In main field over 150 plants are used as host for Sandalwood, some of the good hosts being *Casuarina equisetifolia, Acacia nilotica, Pongamia pinnata, Melia dubia, Wrightia tinctoria and Cassia siamea.*

## Propagation techniques

Seed beds are formed with sand and red soil in 3:1 ratio of and are thoroughly mixed with nematocides (Ekalux or Thimet at 500 g per bed of 10 m × 1 m). Around 2.5 kg seed is needed per bed. Seeds are spread uniformly over the bed and covered with straw which is later removed when the leaves start to appear on the seedlings. Seedling wilt is the major problem in nursery. The initial symptom wilting of leaves followed by chlorosis and root decay leading to death of seedlings. Seed beds are sprayed with fungicide Dithane Z-78 (0.25%) once in 15 days to avoid fungal attack and 0.02% Ekalux solution once in a month to avoid nematode attack. When seedlings have reached 4 to 6 leaf stage, they are carefully removed from beds with all roots intact and transplanted to polybags of 30 × 14 cm size along with a seed of *Cajanus cajan, Alternanthera sessilis, Cassia fistula, Mimosa pudica* etc. or seed of any host plant for better growth of sandal. Polybags should contain a mixture of sand: red earth: farmyard manure in 2:1:1 ratio. Roots of seedlings should not be allowed to dry while transplanting. Shade can be provided immediately after transplantation for nearly a week. Polybags containing seedlings should be watered regularly (once a day) but excess moisture is to be avoided. Seedlings grow to a plantable size of about 30 cm height in 6–8 month time. A well branched seedling with a brown stem is ideal for planting in the field.

## Planting

The pit of 45 cm × 45 cm × 45 cm size preferably at 4 m × 4 m spacing is dug for planting the sandalwood seedlings and is kept distance. Healthy seedlings of sandalwood, preferably above 30 cm in height, are planted in the pits. Miscellaneous secondary forest species are planted in the same pit as host plants, or they may be planted in separate pits in a quincunx pattern.

## Disease

Spike disease caused by mycoplasma–like organisms (MLO) is one of the important diseases of sandalwood. The disease results in extreme reduction in leaf size accompanied by stiffening and reduction of internode length and later the entire shoot have the appearance of a spike inflorescence.

## Irrigation

Though sandalwood is mainly a rainfed crop, young plants require irrigation in dry summer months at 15–20 days interval till they are fully established.

## Heartwood Formation and Oil Content

Heartwood formation generally starts in sandal trees at around 10–13 years of age, but what triggers this process have not been very well understood.

- Young trees (height less than 10 m, girth less than 50 cm, and heartwood diameter 0. 5–2 cm) have heartwood with 0. 2–2% oil content, which has 85% santalol, 5% acetate, and 5% santalenes.
- Mature trees (height 15–20 m, girth 0. 5–1 m, and heartwood diameter 10–20 cm) have heartwood with oil content of 2– 6. 2%, which has over 90% santalol, 3–5% acetate, and 3% santalenes.

## Harvesting

The Sandalwood trees take 30-60 years to reach the age of harvesting. The soft wood is removed first; the hard wood is then chipped and after that milled. The powder is soaked in water for 48 hours before distillation. The process of distillation takes 48 hours to complete. The oil is rectified by re–distillation and filtration.

## Utilization

Sandalwood oil is a viscous liquid pale yellow to yellow in colour, with a sweet and very distinct fragrance which can be warm, woody, spicy, milky and nutty. It is extensively used in aromatherapy and in perfumery, cosmetics, and pharmaceutical industries. The current (2012) cost of sandalwood oil sold at the Karnataka Government outlet is Rs. 1500 for 5 which works out to be Rs 3 lakhs/kg.

## Sandalwood in Agroforestry Systems

When sandalwood is to be planted under agroforestry system, a spacing of 6m × 3m with amla planted in quincuxial design between sandalwood trees appears to be promising. To accommodate higher density of plants, a spacing of 4m × 4m with a long–term host like *Casuarina equisetifolia* in the center, also at the same spacing in quincuxial design of planting is also followed. Spacing of 5m × 4m is also favored by farmers with host plant planted at 2.5m distance within the sandalwood row and sandalwood rows separated at 4m distance to promote inter-cultivation of annual crops. This spacing ensures cultivation of agricultural crops like green gram, cowpea, horse gram, field bean or low spreading legume fodder during the initial years. The cost of raising sandalwood under agroforestry plantations may be marginally higher than raising sandalwood block plantations due to additional intercultural operations. However, this will be compensated by additional returns from field and horticultural crops. Sandalwood plants under agroforestry practices have been observed to perform well in terms of growth parameters possibly due improved soil physico-chemical properties. However, there is a lack of data of sandalwood growth and related heartwood formation under agroforestry situations.

**Fig. 9.1:** Sandalwood based agroforestry system (BAU, Ranchi, Jharkhand and IWST, Bangalore)

## Economics of Sandalwood based agroforestry system

The total cost of cultivation over the 15 years period came around Rs. 19.87 lakhs/ha and the total benefits around Rs. 143 lakhs/ha (IWST, 2008). Of the total cost nearly 30% worked out to be protection costs. The revenue from sandalwood tree extraction and processing in 15 years comes around Rs. 25,000/tree with 3 kg heartwood, 10 kg mixed wood, and 21 kg sapwood per tree, while in 20 years, it was estimated to be Rs. 31,400/tree with 4 kg heartwood, 12 kg mixed wood, and 25 kg sapwood per tree (Viswanath *et al.*, 2010).

In 2017, Karnataka Soaps and Detergent Limited (KSDL) had procured around 75 mature sandalwood trees (18 years age) having a gbh (girth at breast height) ranging from 60–75cm from a farmland in Nelamangala, Bangalore. On an average, the trees yielded 15 kg heartwood and KSDL valued the trees after fixing a rate of Rs. 6400/kg for heartwood, Rs. 2000/kg for mixed wood, Rs. 80/kg for sapwood, and Rs. 30/kg for bark and chips.

**Table 9.1:** B:C ratio analysis of different Sandalwood based agroforestry system in India

| Agro-forestry system | B/C ratio | References |
|---|---|---|
| Sandalwood+ aonla | 2.80 | Viswanath *et al.*, 2010 |
| Sandal+aonla+horsegram | 2.80 | |
| Sandal block plantation | 3.30 | |
| Sandal+gooseberry | 3.80 | APANews, 2009. |
| Sandal+gooseberry+horsegram | 3.80 | |
| Sandal block plantation | 3.40 | |
| Sole Sandalwood | 2.58 | Divakara *et al.*, 2018 |
| Sole Red gram | 1.68 | |
| Sandalwood+ Red gram | 1.93 | |

## Conclusion

Sandalwood farming especially under sandalwood based agro-forestry practices has immense potential in terms of revenue generation and optimal utilization of land and resources.

## References

APANews 2009 Asia-Pacific Agroforestry Newsletter No. 34, Bangkok, Thailand, 11 page.

Divakara BN, Viswanath S, Nikhitha CU and Kumar S 2018 Economics of Santalum album L. Cultivation Under Semiarid Tropics of Karnataka, India. *Forest Research* 7(3) 1–11.

Sandalwood Information Centre 2104 Institute of Wood Science and Technology, Bangalore.

Viswanath S, Dhanya B, Purushothaman S and Rathore TS 2010 Financial viability of sandal based agroforestry practices in Southern India. *Indian journal of Agroforestry* **12**(2) 14–20.

CHAPTER - 10

# Bamboo based Agroforestry System

## BAMBOO

Bamboo belongs to the grass family Poaceae (Gramineae). Around 90 genera and 1200 species of bamboo are known to exist throughout the world (FAO report, 2007). Bamboo is also known as one of the fastest growing plants in the world. The growth rate ranges from 30–100 cm per day in growing season. It can grow to a height of 36 m and diameter of 1–30 cm. In India, bamboos report for around 12.8 percent of the total forest cover which is one of the largest bamboo resources in the world. Bamboo forests occupy an area about 16 million hectares in India where M.P. has maximum bamboo bearing area of 2 m ha followed by Maharashtra (1.54 m ha), Arunachal Pradesh (1.49 m ha) and Odisha (1.18 m ha) (ISFR, 2019). India is second only to China in bamboo production with 3.23 million tons per year (Tewari, 1992). Bamboo is an important non-wood forest product found in forest area as well as outside forest areas in the country. The 'bamboo', also called as poor man's timber, green gold, cradle to coffin timber etc. is one of the most important species having wide range throughout the country except Jammu & Kashmir and Ladakh and has potential to make most important contribution to the rural economy of India. Bamboo also plays an important role in carbon sequestration and biodiversity conservation. On an average, green and litter biomass of bamboos has significantly higher concentration of potassium than other trees (Raizada *et al.*, 2002).

### Botanical Characters

Bamboos are evergreen, monocotyledonous plants which produce primary shoots without any later secondary growth. Every shoot has a distal aerial part called the culm, a proximal, ground-level part known as culm neck, and a bottomless part called the rhizome. Culms contain of nodes and internodes, the former with meristematic tissue from where culm sheaths and branches happen. Rhizomes generate the roots, top growth, and new rhizomes. Then in spring and sometimes fall, they push the accumulated energy into new shoots (culms), which develop to full size in about 30–60 days. The branches and leaves grow in an additional 30–60 days. Shoots of some species have truly been clocked growing 1.20m in 24 hour period. Bamboos are broadly classified into monopodial (or "running") bamboos with "leptomorph" rhizome systems, and sympodial (or "clumping") bamboos with "pachymorph" rhizome systems. The internodes of leptomorph rhizomes are longer than broad, and lateral buds on nodes can create either new shoots or other rhizomes. The internodes of pachymorph rhizomes are broader than long, and lateral buds on nodes create only rhizomes. In sympodial bamboos, new culms develop from buds on stretched out culm necks (pseudo-rhizomes) rather than from buds on rhizomes. Generally, monopodial bamboos are mostly found in temperate climates with cool, wet winters, and sympodial bamboos to tropical climates with a pronounced dry season.

### Area

Around 19.8 million hectares of bamboo reserves are found in India, China and Myanmar which represent 80% of the world's bamboo forests. India represents the global bamboo growing landscape with 136 species, 23 genera spreading over 16 million ha; which makes it, the second largest bamboo growing country in the world (ISFR, 2019). But ironically India's share in global bamboo trade and commerce is only 4% though it possesses 45% of global bamboo growth. Salam, 2013 reported that, there has been a broad gap in the demand and supply scenario so far as the bamboo based resources are concerned. The total demand for bamboo products has been estimated at 26.9 million tons against a supply scenario of 13.47 million tons.

### Distribution

Two-third of the growing stock of bamboos in the country is available in the North–Eastern states. They abundantly occur in Andhra Pradesh, Arunachal Pradesh, Assam, Manipur, Meghalaya, Mizoram, Nagaland, Sikkim, Tripura, Orissa, West Bengal and Madhya Pradesh states. Some species are also found scattered in other parts of the country both in the hills and the plains.

### Climate requirement

Bamboo occurs between the latitudinal range of 46 °N and 47 °S and altitudinal range of 0–4000 m from the mean sea level covering tropical, subtropical, temperate and alpine regions (Dransfield, 1992). It is found to grow practically all over the country except Jammu & Kashmir, particularly in the tropical, sub-tropical and temperate regions where the annual rainfall ranges between 1,200 to 4,000 mm and the temperature varies between 16 °C and 38 °C.

### Soil

Bamboos can be grown and cultivated on a broad variety of soils except on rock-strewn soils. Bamboo plantation requires well drained sandy soil to clay soil that should have a pH range of 4.5–6.0. If we talk about India, then the Barack Valley area is best suited for bamboo cultivation due to best soil and ideal climate conditions.

## Variety

India is home to about 125 indigenous and 11 exotic species of bamboo from 23 genera. The principal bamboo genera occurring in India are Arundinaria, Bambusa, Chimonobambusa, Dendrocalamus, Dinochola, Gigantochloa etc. *Dendrocalamus strictus* has major share of 45 percent followed by *Bambusa bambos* (13%), *D. hamiltonii* (7%), *B. tulda* (5%), *B. pallida* (4%) and rest other species (6%).

**Table 10.1:** Recommendations of bamboo species for different agro-ecological zones in India.

| S. No | Name of the Centre | Ecological Zone | Species showing superior growth |
|---|---|---|---|
| 1. | Allahabad Agriculture Institute (Deemed University), Naini, Allahabad, Uttar Pradesh | Indo-gangetic | *B. balcooa, B. nutans, B. tulda, B. vulgaris* |
| 2. | Bidhan Chandra Krishi Vishwa Vidyalaya, Nadia, West Bengal | East Zone | *B. balcooa, B. bambos, B. tulda, B. nutans,* |
| 3. | G.B. Pant University of Agriculture and Technology, Pantnagar, Uttarakhand | Tarai (Himalayan foot hill) | *D. hamiltonii, B. balcooa, B. nutans, B. vulgaris* |
| 4. | Institute of Himalayan Bioresource Technology, Palampur, H.P. | Temperate zone | *D. hamiltonii, B. balcooa, B. nutans, B. tulda, B. bambos* |
| 5. | Institute of Wood Science and Technology, Bangalore | Dry Tropical | *D. hamiltonii, B. tulda, B. balcooa* |
| 6. | Kerala Forest Research Institute, Peechi, Kerala | Wet Humid Tropic | *B. balcooa, B. nutans, D. hamiltonii, B. bambos* |
| 7. | Institute of Forest Genetics and Tree Breeding, Coimbatore, Tamil Nadu | Dry Peninsular | *B. nutans, B. vulgaris, B. balcooa, B. bambos* |

## Propagation techniques

i. **Through Seed**– Bamboo seeds germinate in rainy season after gregarious flowering. The seedlings spring up and survive in large numbers on bare ground. A few of the seedlings develop into clumps after 5 to 10 years.

ii. **Offset Method**– An offset is the lower part of a single culm (usually with 3–5 nodes *i.e.* about l to 2.5 m) with the rhizome axis basal to it and its roots. The culm should be between 1–2 years old and from a well clump. Generally, the method applies to thick walled clump forming species. The culm is cut with a slanting cut and the rhizome to which it is attached is dug up and cut off to a suitable length to include well developed buds. Offsets are usually obtained and transplanted immediately before the rainy season. It is suitable to seal the slant cut of the culm with wax or cow dung or earth.

iii. **Rhizome Method**– The 'rhizome' is an underground stem consisting of two parts, the rhizome proper and the rhizome neck which is the base of the rhizome proper. As a stem, the rhizome, has nodes. The rhizome has good roots or root primordia and buds at all or most of the nodes. Rhizome cuttings are sections of fresh living rhizomes of the preceding year measuring about 15 to 30 cm long, containing at least one bud are planted in pits.

iv. **Stem or Culm Cuttings**– For culm–or stem–cutting propagation technique, generally, culm segments of 1 or usually 2–3 nodes bearing healthy buds or branches are used. Propagation through culm cuttings is possible throughout the year with varying success with or without growth hormone if watering and temperature is maintained with a thick pad of straw mulching. However, the best effect so far obtained with two node culm segments of 8–21 months old during March–May without rooting hormones in the tropical situations of eastern part of India. Propagation through branch cuttings are also possible with poor rooting percentages and rhizome development.

v. **Layering Method**– Layering is a method of bringing a culm or branch in contact with soil so that propagation occurs. There are 4 types of layering– i) ground or simple layering ii) stump layering iii) air layering or marcotting and iv) seedling layering. For ground layering and stump layering appropriate media are 1:1 coconut husk : sand, cow dung : sand, sawdust : sand or peat moss : sawdust. For air layering coconut fibre or coir or peat moss is most suitable.

**Table 10.2:** Nutrient management of Bamboo

In case of bamboo plantation, fertilizers were applied in g/plant.

| Year | Fertilizers | RDF(g/plant) |
|---|---|---|
| I | N : P : K | 50 : 25 : 25 |
| II | N : P : K | 100 : 50 : 50 |
| III | N : P : K | 150 : 75 : 75 |

### Irrigation

Irrigation has to be done regularly as the bamboos are grown on nursery beds. Water should be provided immediately after transplanting seedlings from nursery to field. Bamboo plants are very sensitive to water logging hence soil must be drained out especially during heavy rainfall or flooding areas. Drip irrigation system can also be adopted for better utilization of water.

### Harvesting

Harvesting of bamboo can be started from the fifth year onwards and in case of commercial farming, harvesting must be done from the sixth year. The unit cost for one hectare of bamboo plantation is around Rs. 23500 spread over a period of five years. In the first harvest in sixth year, 6 culms can be harvested followed by 7 culms in the seventh year and so on.

The yield and income generation from bamboo plantation, increases each year starting from the sixth year. Hence we can say that bamboo is a cash crop having short rotation period, faster growth and gives recurring economic returns, generation after generation.

## Utilization

Bamboo represents a society of woody perennial grass that occurs in the tropical and subtropical evergreen and deciduous forest area of Asia–Pacific region. The major utilization of bamboo resources include paper and pulp industry, fuel, food, feed, house construction, rafting, handicraft, umbrella handle and scaffolding sectors, producers of several articles of domestic applications, as well its role a significant provider of a host of ecosystem services.

## Bamboo in Agroforestry System

The integrated approaches for enhancing bamboo production through various strategies including positioning of agroforestry in the production context is expected bridge the demand and supply gap in domestic market as well as to encourage inclusive development in the context of climate change with multiplayer impacts on employment generation, income generation and food security. Though agroforestry based bamboo production models or systems help to bridge the supply gap in the bamboo trade and market to certain extent, there need to be further solid strides taken in the areas of market development, integration, value addition, value chain management, skill development, insurance etc in the production and utilization domains.

**Table 10.3:** Bamboo based agroforestry systems in India under AICRP on agroforestry.

| Agroclimatic Zone | Research Centre | Bamboo Species | Possible Intercrops |
|---|---|---|---|
| Central Plateau and Hills Region | CAFRI Jhansi | *D. Strictus* | Gram, lentil and sesame |
| Eastern Himalayan Region | AICRP–AF, Guwahati | *B. Tulda* and *B. balcooa* | Ginger, turmeric, maize, mustard, chili, colocasia, sesamum, niger, turmeric, black gram, groundnut |
| Eastern Plateau & Hills Region | AICRP–AF, Jhargram | *B. Tulda* and *B. balcooa* | Chili, mustard, sesamum, niger, turmeric, black gram, groundnut |
| Southern Plateau & Hills Region | AICFRP–AF, Dharwad | *B. bambos* | Soyabean, mustard, sesamum, greengram, black gram, cotton, Arhar, Horse gram, Sesamum |

| Agroclimatic Zone | Research Centre | Bamboo Species | Possible Intercrops |
|---|---|---|---|
| East Coast Plains & Hills Region | AICRP–AF, Bhubneswar | *D. Strictus* | Arhar, Horse gram, sesamum, colocasia, cowpea, turmeric |
| West Coast Plains & Gats Region | AICRP–AF, Dapoli | *O. stocksii* | Finger millets, green gram, blackgram, pigeon oea, niger |

In India, cultivation of bamboo is done since ancient times at homesteads, small scale farm lands and farm or field boundaries for livelihood support, various purposes including as live fence to protect agricultural crops like windbreak. They are managed as grooves or following several agroforestry models. Different bamboo based agroforestry systems like agrisilviculture, agrihorticulture, silvipasture etc. are prevalent in different states namely Assam, Arunachal Pradesh, Sikkim, Kerala, Tamilnadu, Chhattishgarh, Orissa, Jharkhand and Bihar.

**Table 10.4:** Some successful Agroforestry Systems with Bamboo in India (*Source:* Nath *et al.*, 2009).

| Sl | Region/state | Agricultural Components | Forestry Components | Source |
|---|---|---|---|---|
| | | AGRICULTURAL CROPS : | | |
| 1 | Tamil Nadu | Soybean (*Glycine max*) | *D. strictus* | Seshadhri, 1985 |
| 2 | Konkan, Karnataka | Mango (*Mangifera indica*), Cashewnut (*Anacardium occidentale*), Jackfruit (*Artocarpus heterophyllous*), Kokum(*Garcinia indica)* and Rubber (*Ficus elastic*) | *B. bambos* | Wagh and Rajput, 1991 |
| 3 | Central Region | Wheat (*T. aestivum*), rice (*Oryza sativa*), maize (*Zea mays*), jowar (*Sorghum bicolor*), bajra (*Pennisetum glaucum*), pulses, oil seeds etc. | *Bamboos* of different spp. | Dwivedi, 1994 |
| 5 | Southern Region | Rice (*Oryza sativa*), tobacco (*Nicotiana tabacum*), Chillies (*Capsicum frutescens*) and Sugarcane (*Saccharum officinarum*) | *Dendrocalamus hamiltonii* | Dwivedi, 1994 |
| 6 | N E Region | Paddy (*Oryza sativa*) | *D. hamiltonii* | Dwivedi, 1994 |

| Sl | Region/state | Agricultural Components | Forestry Components | Source |
|---|---|---|---|---|
| 7 | Jabalpur, M.P. (Degraded Soil) | Soybean (*Glycine max*), Niger (*Guizotia abyssinica*), Moong (*Phaseolus aureus*), Wheat (*Triticum aestivum*), Urad (*Phaseolus mungo*), Pegion pea (*Cajanas cajan*) and Mustard (*Brassica campestris*). | *B. bamboos, B. nutans* and *D. strictus* | Behari, 2001 |
| 8 | Kallipatty, Tamil Nadu | Pigeon pea(*Cajanas cajan*, Soybean(*Glycine max)* and Turmeric(*Curcuma longa*) | *B. bamboos* | Shanmughavel and Francis, 2001–02 |
| 9 | N–E India (Degraded Soil) | Ginger (*Zingiber officinale*) | Edible | Jha and Lalnunmawia, 2004 |
| 10 | Coorg, Karnataka | Ginger (*Zingiber officinale*) | *D. brandisii* | Viswanath, Dhanya, and Rathore, 2007 |

## Economics of bamboo based agroforestry system

Returns from sole bamboo were very low because of under–utilization of land resource, whereas under agroforestry, agriculture/horticulture/fodder crops were raised along with bamboo in the interspaces, which enhanced the overall productivity of the system. Kaushik *et al.*, 2015, reported that in general agroforestry systems, crop yields are reduced due to competition of above and below ground resources and shading effect, but agroforestry systems gives overall higher economic returns due to supplemental timber/wood yield from tree components. The studies done in the multi tier home gardens of Kerala, shows that bamboo (*Bambusa bambos*) holds the second position in terms of income generation and benefit–cost ratio (BCR), among the crop groups. The higher BCR ratio of bamboo was due to very low inputs and high farm price (Krishnankutty, 2004). Behari (2001) studies that, developed successful seven agroforestry models with three bamboos (*B. bamboos, B. nutans and D. strictus*). The intercrops grown were: Soybean, Niger, Moong, Wheat, Urad, Pigeon and Mustard. These models were developed to re–establish the degraded agricultural lands in central part of India. The studies carried out in the abandoned paddy fields in Coorg, Karnataka, planting bamboo (*D. brandisii*) at 6m × 6m spacing and intercropped with ginger showed the highest NPV (net present value) and LEV (Land expectation value) are found. This may be attributed to low input costs integrated with bamboo farming and higher market value of the produce over a longer period.

**Fig. 10.1:** Bamboo based agroforestry system (BAU, Ranchi, Jharkhand)

Seshadri (1985) observed that growing of soybean as an intercrop with bamboo during the first six years is technically feasible and economically viable agroforestry modes. He also reported that the period of intercropping can be extended further in wider spacing of the bamboo culms and judicious management of the bamboo canopy.

Balaji (1991) reported that the scope of bamboo based agroforestry system is very wide because of the unsure weather condition and increasing cost of labour concerned in agriculture these days. In study on systematic bamboo plantation intercropped with mango, cashew nut, jack fruit, kokum and rubber in the konkan region of Karnataka, bamboo is reported to be the most profitable or income generation among the agriculture crops studied and cashew nut and mango ranked next to bamboo was found (Wagh and Rajput, 1991).

**Table 10.5:** B:C ratio analysis of different bamboo based agroforestry system in India

| Agro-forestry system | B/C ratio | References |
|---|---|---|
| *B. tulda* 10×10 + Rice | 0.50 | Dhara, P.K. and Banerjee, H. 2007 |
| *B. tulda* 12×10 + Rice | 0.31 | |
| *B. tulda* 10×10 + Pigeon pea | 2.03 | |
| *B. tulda* 12×10 + Pigeon pea | 2.02 | |
| *B. tulda* 10×10 + Bottle gourd | 2.28 | |
| *B. tulda* 12×10 + Bottle gourd | 2.13 | |
| *B. tulda* 10×10 + Turmeric | 0.70 | |
| *B. tulda* 12×10 + Turmeric | 0.51 | |
| *B. balcooa* 10×10 + Rice | 0.24 | |
| *B. balcooa* 12×10 + Rice | 0.23 | |
| *B. balcooa* 10×10 + Pigeon pea | 1.83 | |
| *B. balcooa* 12×10 + Pigeon pea | 1.99 | |
| *B. balcooa* 10×10 + Bottle gourd | 2.04 | |
| *B. balcooa* 12×10 + Bottle gourd | 1.99 | |
| *B. balcooa* 10×10 + Turmeric | 0.44 | |
| *B. balcooa* 12×10 + Turmeric | 0.43 | |
| Sole *B. tulda* 10×10 | 0.72 | |
| Sole *B. tulda* 12×10 | 0.63 | |
| Sole *B. balcooa* 10×10 | 0.36 | |
| Sole *B. balcooa* 12×10 | 0.18 | |
| Sole Rice | 0.18 | |
| Sole Pigeon pea | 4.33 | |
| Sole Bottle gourd | 4.78 | |
| Sole Turmeric | 0.59 | |
| Bamboo 10×10 +Cowpea (pod) | 1.39 | Research Station, O.U.A.T., Bhubaneswar, 2010 |
| Bamboo 12×10 +Cowpea (pod) | 1.34 | |
| Bamboo 10×10 +Blackgram (seed) | 1.67 | |
| Bamboo 12×10 +Blackgram (seed) | 1.65 | |
| Bamboo 10×10 +Greengram (seed) | 1.57 | |
| Bamboo 12×10 +Greengram (seed) | 1.50 | |
| Bamboo 10×10 +Sesamum (seed) | 1.53 | |
| Bamboo 12×10 +Sesamum (seed) | 1.49 | |
| A1B1 | 2.36 | Ilorkar *et al.,* 2016. |
| A1B2 | 2.41 | |
| A1B3 | 2.80 | |
| A1B4 | 1.90 | |
| A2B1 | 2.36 | |
| A2B2 | 2.56 | |
| A2B3 | 2.30 | |
| A2B4 | 2.00 | |

| Agro-forestry system | B/C ratio | References |
|---|---|---|
| A3B1 | 2.70 | |
| A3B2 | 2.70 | |
| A3B3 | 3.06 | |
| A3B4 | 1.93 | |
| A4B1 | 2.47 | |
| A4B2 | 2.18 | |
| A4B3 | 2.19 | |
| A4B4 | 2.10 | |
| Where, A– Conservation practices<br>A1– CCT<br>A2– CCT + Sloping trench<br>A3– Circular trench round the clump<br>A4– control<br>Sub treatments:---------<br>B– Felling intensity<br>B1– 10% felling of Bamboo<br>B2– 20% felling of Bamboo<br>B3– 30% felling of Bamboo<br>B4–Control | | |
| Sole *B. balcooa* (after 5 years) | 1.81 | Kumar *et al.*, 2016 |
| Sole *B. vulgaris* | 1.17 | |
| Sole *D. strictus* | 1.09 | |
| Bamboo + sesame (10×10 m) | 2.83 | Dev *et al.*, 2017 |
| Bamboo + sesame (12×10 m) | 2.59 | |
| Sole sesame | 1.43 | |
| Sole bamboo (10×10 m) | 3.52 | |
| Sole bamboo (12×10 m) | 3.20 | |
| *B. vulgaris* + Greengram | 2.60 | Shudhagar *et al.*, 2017 |
| *B. vulgaris* + Blackgram | 3.10 | |
| *B. vulgaris* + Cowpea | 3.00 | |
| *B. vulgaris* + Gingelly | 1.20 | |
| *B. vulgaris* + Bhendi | 1.90 | |
| *B. vulgaris* + Brinjal | 2.00 | |
| *B. vulgaris* + Sudan grass | 4.40 | Puran *et al.*, 2018 |
| *B. vulgaris* + Napier grass | 2.60 | |
| *B. vulgaris* + Guinea grass | 3.74 | |
| Sole *B. vulgaris* | 2.67 | |
| Sole Sudan grass | 3.70 | |
| Sole Napier grass | 0.72 | |
| Sole Guinea grass | 2.11 | |

| Agro-forestry system | B/C ratio | References |
|---|---|---|
| 10 m × 10 m Bamboo + sesame-chickpea | 2.6 | Dev *et al.*, 2020 |
| 12 m × 10 m Bamboo + sesame-chickpea | 2.7 | |
| 10 m × 10 m Sole bamboo | 6.0 | |
| 12 m × 10 m Sole bamboo | 5.4 | |
| Sole crop (sesame-chickpea) | 2.7 | |

## Conclusion

Government of India should promote bamboo based agroforestry system for better economic returns to the farmers in semi-arid tropic region under drought and unsure weather conditions. The economic impact of the bamboo based agroforestry system significantly influences general economic development and this system is especially important as well as significant for developing country like India or other. Under this system because of growing of various intercrops, products are obtained even in the early stages of plantations and the income is much higher than any other system of cropping. Bamboo species and horticulture crop, fodder crop, medicinal and aromatic plant, agricultural crops like paddy (upland), pigeon pea, bottle gourd and turmerich as immense potentiality of providing livelihood security to the poor farmers of India through self employment and higher income. There is still a substantial need to promote bamboo-based agroforestry systems as well as utilization of bamboos to the extent possible in wasteland particularly in economic.

## Reference

Balaji S 1991 Agroforestry for prosperity. *Forest News* **1**(3) 9–11.

Behari B 2001 Agroforestry models of bamboo cultivation on degraded agricultural lands. Ph.D. thesis. Guru Ghasidas University, Chhattisgarh.

Dev I, Ram A, Ahlawat SP, Palsaniya DR, Singh R, Dhyani SK, Kumar N, Tewari RK, Singh M, Babanna SK, Newaj R, Dwivedi RP, Kumar RV, Yadav RS, Chand L, Kumar D and Prasad J 2020 Bamboo–based agroforestry system (*Dendrocalamus strictus* + sesame–chickpea) for enhancing productivity in semi-arid tropics of central India. *Agroforest System* 1–15 pages.

Dev I, Ram A, Ahlawat SP, Palsaniya DR, Newaj R, Tewari RK, Singh R, Sridhar KB, Dwivedi RP, Srivastava M, Chaturvedi OP, Kumar RV and Yadav RS 2017 Bamboo (*Dendrocalamus strictus*) + sesame (*Sesamum indicum)* based agroforestry model: A sustainable livelihood option for farmers of semi-arid region. *Indian Journal of Agricultural Sciences* **87**(11) 1528–1534.

Dhara PK and Banerjee H 2007 Development of Bamboo–based Agroforestry Systems for Six Agro-climatic Zones. Regional Research Station, (Red & Laterite Zone), Bidhan Chandra Krishi Viswavidyalaya, Jhargram–721507, Dist. Paschim Medinipur, West Bengal, India.

Dransfield S 1992 The Bamboos of Sabah. Sabah Forest Report, No. 14. Malaysia.

Ilorkar VM, Kausadikar PH and Khobragade YR 2016 Studies on effect of Water Conservation Practices and Felling Intensity on production of Bamboo (*Denrocalmus strictus*). National Symposium on Agroforestry for Environmental Challenges, sustainable land use, biodiversity conservation and rural livelihood options. 3–5 December 2016, ICAR -CAFRI, Jhansi.

ISFR 2019 Indian State of Forest Report, Chapter–8 Bamboo resource of the country. Forest Survey of India (Ministry of Environment Forest and Climate Change), Dehradun, Uttarakhand, India, 123–134 pages.

Kaushik N, Deswal RPS and Mali S 2015 Agroforestry systems for fodder production under rainfed conditions. *Eco. Env. and Cons.* **21**(2) 901–904.

Krishnankutty CN 2004 Benefit–cost analysis of bamboo in comparison with other crops in mixed cropping home gardens in Kerala State, India, *J. Bamboo and Rattan* **3**(2) 99–106.

Kumar H, Mehera B, Lal SB and Umrao R 2016 The potential of bamboo cultivation as a way forward in Improving livelihood: a case study

Nath S, Das R, Chandra R and Sinha A, 2009 Bamboo Based Agroforestry for Marginal Lands with Special Reference to Productivity, Market Trend and Economy, symposium on 21 March 2009, Ranchi, Jharkhand.

Puran A, Malik MS, Oroan PR and Kumar A 2018 Study on Growth and Economics of Bamboo Based Silvipastoral System. *Int.J.Curr.Microbiol.App.Sci.* **SI7** 4273–4277

Raizada A, Singh C and Singh G 2002 Production, nutrient dynamics and breakdown of leaf litter in six forest plantations raised on gravelly flood plain in the lower Western Himalayas. *J. Trop. For. Sci.* **14**(4) 499–512.

Research Station 2009 Development of Bamboo–based Agroforestry Systems for Six Agro-climatic Zones. O.U.A.T., Bhubaneswar, Odisha, 131 page.

Salam K 2013 Connecting the poor: Bamboo, Problems and prospect. South Asia Bamboo Foundation (SABF) retrieved 17 December 2013 from jeevika. org/bamboo/2g–article–for–nbda.docx.

Seshadri P 1985 Intercropping of bamboo (*D. strictus*) with Soybean–An agroforestry study. Ph.D. thesis, Tamil Nadu Agricultural University, Coimbatore, 480 page.

Shudhagar RJ, Divya MP, Ramah K and Sivakumar K 2017 Pulpwood species based agroforestry models. In book Parthiban KT and Seenivasan R plantation and agroforestry–pulpwood valve chain approach, Scientific Publishers (India), 291 page. ISBN–9789386347985

Tewari DN 1992 A monograph of Bamboo. International Book Distributors, Dehradun, India, 495 page.

Wagh R and Rajput JC 1991 Comparative Performance of Bamboo with the Horticultural Crops in Konkan. In: IV International Bamboo Workshop on Bamboo in Asia and Pacific, Chiangmai, 27–30 November 1991, Thailand, International Development Research Centre, 5–86 pages.

Chapter - 11

# Mango based Agroforestry System

## MANGO

Mango is the National fruit of India and Pakistan and national tree of Bangladesh. It is also called King of fruit and Bathroom Fruit. Its botanical name is *Mangifera indica* which comes under the family Anacardiaceae. Its origin is considered to be South East Asia or Indo-Burman region. Mango is being cultivated in southern Asia for nearly six thousand years. Mango is one of the most important and widely cultivated tropical fruits of the world. India ranks first among world's mango producing countries accounting for about 54.7 % of the world's mango production. Other major mango producing countries include China, Thailand, Mexico, Pakistan, Philippines, Indonesia, Brazil, Nigeria and Egypt. In India, it is grown in an area of 21.63 lakh hectares with an annual production of 185.27 lakh tones (NHB, 2016). In India Uttar Pradesh is the leading state in mango both in area and production.

Ripe mango is rich source of vitamin A while a good amount of vitamin C is present in unripe mango. Yellow colour of mango pulp is due to beta-carotene. Ripe mango is generally eaten as fresh fruit, besides there are various other culinary uses. Sour unripe mangoes are used to prepare chutney, amchur and pickle. Mango is widely used in fruit processing industry and contributes maximum part of export from India.

### Botanical Description

Mango tree is an evergreen tree of varying size and shape. It has a deep taproot and profuse surface roots, a stout trunk and an umbrella-shaped crown that may reach 10–40 m high. Leaves are simple, alternate, borne on 1–12.5 cm long petioles. Length of leaves is 16–30 cm on flowering branches and up to 50 cm long on sterile branches while width 3–7 cm. Young leaves are orange-red and turn shiny dark green on the upper surface when they mature. Mango tree flowers are fragrant, pentameric, greenish-white or pinkish. The fruit is a large fleshy drupe of variable size, shape, colour and taste, with a woody endocarp (the pit), a resinous edible mesocarp and a thick exocarp, the peel. Green when unripe, the fruit turns orange-reddish as it ripens. The fruit takes between 3 and 6 months to ripen.

Chromosome number:– 2n=4x=40 (Allopolyploid)

Type of fruits:– Drupe

Type of inflorescence:– Panicle

Edible part:– Mesocarp

### Climatic requirement

Mango is primarily a tropical climate tree but it also cultivated in subtropical area. It thrives well in almost all the regions of the country but cannot be grown commercially in areas above 600 m. It cannot stand severe frost, especially when the tree is young. High temperature by itself is not so injurious to mango, but in combination with low humidity and heavy winds, it affects the tree adversely. Optimum temperature for growth is 24 to 28 °C.

Mango varieties usually thrive well in places receiving rainfall in the range of 1000 mm but can survive 750–3750 mm /annum rainfall. The distribution of rainfall is more important than its amount. Dry weather before blossoming is conducive to profuse flowering. Rain during flowering is detrimental to the crop as it interferes with pollination. However, rain during fruit development is good but heavy rains cause damage to ripening fruits. Strong winds and cyclones during fruiting season can play havoc as they cause excessive fruit drop.

### Soil

Mango can be grown on a variety of soils but alluvial and sandy loam soils are most suitable. The most desirable soils for mango should be of medium texture, deep (2 to 2.5 m) well drained with low water table (below 180 cm in all seasons) and have a pH range of 6.0 to 7.5. Mango cannot tolerate high soil salt content (not more than 0.05 %).

### Cultivation technology

Mango is commercially propagated by Veneer grafting. One or two year old veneer grafts should be collected from an authentic source. The grafts should have a minimum of 4 whorls of leaves with strong graft union.

### Different propagation methods:

- Veneer grafting– It is commercial propagation method of mango. Suitable time for veneer grafting is August.
- Stone grafting– Best time of stone grafting is September. It is practices in Konkan region of Maharashtra.
- Inarching– This method is adopted by East Indian farmers. Best time for Inarching is July.

### Root stock

- Vellaicollumban is allo-octaploid (2n=2x=80) and it is a dwarf root stock.
- Other dwarfing root stock– Kurkan, Olour, Nileswara Dwarf and Totapuri Red Small.
- Alkali soil tolerance rootstock:– Kurkan, Nickerie and Olour.

### Planting

Planting of mango should be done at the beginning of the monsoon (June– July). In areas with heavy rainfall planting should be done at the end of rainy season. The pit of size 1m x 1m x 1m should be dug and filled with top soil, well mixed with 25 kg farmyard manure and 2 kg superphosphate and 100 gm of Folidol powder to avoid termite damage. Planting distances can be around 7 to 10 meters depending on soil depth and vigour of the variety planted. Even 12 meters spacing may be necessary in very deep and fertile soils. The graft should be planted with ball of earth and press the soil all around the base. The graft joint should be above ground level. Staking is done to prevent wind damage. Root-stock sprouts below the graft joint should be removed.

Size of pit– 1m × 1m × 1m

Planting distance– 10m × 10m in dry and 12m × 12m in moist zone

Time of planting– June–July

Flowering time– Feb–March

Fruiting time:– May–July

Maturity– 90–110 days after flowering fruit is ready for harvesting.

Yield– 90–120kg/tree or 15–20t/ha

### Pruning

Pruning is done to remove criss- cross branches may be done so that center of the tree is opened out and inner branches are exposed to sunlight. Pruning of the dried twigs and branches should be done with pruning saw during June–July. Pruning in mango encourages production of new shoots.

### Manures and fertilizers

Fertilizers may be applied in two split doses, one half immediately after the harvesting of fruits in June/July and the other half in October, in both young and old orchards followed by irrigation if there are no rains. Foliar application of 3 % urea in sandy soils is recommended before flowering.

The following table gives the details of fertilizer applied (depending upon the age of the plants):

**Table 11.1:** Nutrient Management in Mango

| Age of the plant (in years) | Fertilizer applied |
|---|---|
| 1 | 100g. N, 50g. $P_2O_5$, 100g. $K_2O$ |
| 10 | 1kg. N, 500g. $P_2O_5$, 1kg. $K_2O$ |

### Irrigation

Mango responds well to irrigation, particularly at fruit setting and developmental phases. The young plants at bearing stage should be irrigated frequently. In bearing trees, for obtaining good flowering, irrigation must be stopped at least 2 months before flowering period. Stopping of irrigation creates stress and encourages flower bud formation.

Irrigation should be given at the time of fruit setting and thereafter at regular intervals during fruit development period, up to full development stage. For better quality fruit, irrigation should be stopped 20–30 days before maturity/harvesting the crop. As far as possible, irrigation should be given with drip system. Drip irrigation not only conservers water but also increases water use efficiency. For conservation of rain water, rows (in rectangular system of planting) should be made along the contour in sloppy locations.

**Table 11.2:** Irrigation schedule in Mango

| Age of the plant (in years)/Growth stage | Irrigation schedule |
|---|---|
| 1 | Irrigated at an interval of 2–3 days during dry season. |
| 2–5 | Irrigation interval– 4–5 days . |
| 5–8/ fruit set to maturity | Irrigated after every 10–15 days |
| Full bearing stage | 2–3 irrigations after fruit set. |

## Varieties

India has the richest collection of mango cultivars.

1. Alphonso: Also called Hapus. One of the most popular variety of India, it is mainly grown in Ratnagiri area of Maharashtra and to a small extent parts of south Gujarat and Karnataka. It is susceptible to spongy tissue.
2. Baiganaphalli: It is a widely cultivated, early maturing mango of south India. It is the main commercial variety of Andhra Pradesh.
3. Bombay Green: It is one of the earliest varieties of north India. Its fruits have highest vitamin C content. Highly suitable to Mango-malformation.
4. Chausa: Late-maturing variety of south India, it matures during July or beginning of August. It is the sweetest variety of mango.
5. Dashehari: One of the most popular variety of north India
6. Fazli: Late-maturing variety.
7. Kesar: Popular in Saurashtra region of Gujarat, Kesar is an irregular bearing mango. Fruits ripen to attractive apricot yellow color with red blush. It has good processing quality.
8. Langra: An important commercial mango variety of north India, it is biennial-bearer and a mid-season variety. It is highly prone to fruit dropping.
9. Neelum: It has regular bearing habit.
10. Madhulika: It is an off season cultivar.

11. Niranjan: It is an off season cultivar.
12. MOCH–1: It is an off season cultivar.
13. Lal Sinduri: Powdery mildew resistant variety.
14. Rosica: Mutant variety.
15. Rumani: Apple Shape variety.

**Table 11.3:** Hybrid varieties of Mango

| Hybrid | Place of research | Parentage | Importance characters |
|---|---|---|---|
| Mallika | IARI, New Delhi | Neelum × Dashehari | Highest Vit–A content |
| Amrapali | IARI, New Delhi | Dashehare × Neelum | Dwarf variety of mango. Planting Distance: 2.5m × 2.5m (1600 plant planted in 1 ha) |
| Ratna | FRS Vengurla | Neelum × Alphonso | Free from spongy tissue |
| Sindhu | FRS Vengurla | Ratna × Alphonso | Regular- bearer |
| Arka Puneet | IIHR, Banalore | Alphonso × Banganapalli | Free from spongy tissue |
| Arka Aruna | IIHR, Banalore | Baganpalli × Alphonso | Free from spongy tissue |
| Arka Anmol | IIHR, Banalore | Alphonso × Janardhan | Free from spongy tissue |
| Arka Neelkiran | IIHR, Banalore | Alphonso × Neelum | Free from spongy tissue |
| Ambika/CISH–1 | CISH, Lucknow | Amrapali × Janardhan | – |
| Pusa Arunima | – | Amrapali × Sensation | |
| Pusa Lalima | – | Amrapali × Sensation | – |
| Pusa Pratibha | – | Amrapali × Sensation | – |
| Sai Sugandha | – | Totapuri × Keser | – |
| Prabha Sankar | – | Bombay × Kalapady | – |
| Sunder Langra | – | Langra × Sunder Prasad | – |
| Neeleshan | – | Neelum × Baneshan | – |
| PKM–2 | – | Neelum × Mulgoa | – |

## Physiological disorders of Mango

1. Spongy Tissue: It is the major problem in Alphonso. It is a physiological disorder in which fruit pulp remains unripe because of unhydrolyzed starch due to physiological and biochemical disturbances because of high temperature, convective heat and post-harvest exposure to sunlight.

   Management: Use of mulch and post-harvest exposure to low temperatures between 10–15°C for 10–18 hours has been useful in reducing the malady.

2. Biennial Bearing/ Alternate Bearing: Mango tree heavily bear fruits in one year (on year) and a much reduced crop in the following year (off year). Totapari, Neelum, and hybrids in which Neelum is involved as one of the parents are regular bearers.

3. Fruit Drop: In mango, there is a heavy drop of hermaphrodite flowers and young fruits amounting to 99 % or more. The commercially grown varieties, Langra and Dashehari are more susceptible to drop in the month first three weeks of April.

   Management: Regular irrigation during fruit setting and development period can reduce this problem. Application of plant bio-regulators like NAA 25 ppm and 2,4 D 10–15 ppm at 6 weeks after fruit set reduce fruit drop considerably.

4. Black Tip: Black tip is a serious disorder, particularly in the cultivar Dashehari.

5. Mango Malformation: It was first reported in Darbhanga, Bihar in 1891. Malformation is widely prevalent in northern India, particularly in the states of Punjab, Delhi and western U.P. where more than 50 % of the trees suffer from this malady. Mango malformation is of two types viz. vegetative and floral.

   Management: Spraying of Planofix (200 ppm) during the first week of October. Cutting of the malformed twig alongwith the approximate 15cm healthy portion after complete fruit set in May.

## Yield

Graft plants start bearing at the age of 2–3 years (10–20 fruits) to give optimum crop from 10–15$^{th}$ year which continues to increase upto the age of 40 years under good management. Yield of fruits varies considerably according to the variety, climatic conditions, plant population etc. On an average, the yield ranges from 12–22 tons/ha. Mango wood is known as hardwood and it takes 15 to 20 years to grow and ready to use for furniture.

## Utilization

Mango used as mainly edible fruit, sweet, juice, candy, pickle and various medicinal purposes. It also used in road side plantation, windbreak, shelterbelt and aesthetic value. Mango tree leave can be used as fodder, either as cut-and-carry or browsed. Mango wood use for furniture, fuel wood, handicraft, Flooring, Paneling, Veneer,

Plywood, Musical instruments (especially ukuleles, drums, and guitars), Table boards, Beds etc.

## Mango in Agroforestry system

The mango plants when planted at a spacing of 10 m × 10 m provide an ample scope for growing of short duration crops (Swain, 2014). Interspace between the rows of trees can be successfully utilized for cultivation of agricultural crops, fruits, vegetable, spices, medicinal and aromatic plants or fodder crops. Mango based agroforestry is economically viable and more profitable than many of crop rotations followed in the study area. This land use system is also capable for providing employment opportunities of farmers. Mango based agroforestry generates additional employment and income for farm families and improves the socio-economic status of growers. This practice, besides providing income to the grower, suppresses weed growth, improves soil fertility and prevents leaching of nutrients and erosion of soil, achieves higher gross return per unit area.

Intercropping can be taken up till the mango trees attain suitable height and develop canopy (at 5–6 years of age). Leguminous crops like green gram, black gram, gram etc., cereals like wheat, oilseeds like mustard, sesame and groundnut, vegetable crops such as cabbage, cauliflower, tomato, potato, brinjal, cucumber, pumpkin, bitter gourd, apple gourd, lady's finger etc. and spices like chillies can be grown as intercrops. The partial shade loving crops like pineapple, ginger, turmeric etc. can be cultivated in fully grown orchards. In addition to field crops, some short duration, less exhaustive and dwarf type inter- fillers like papaya, guava, peach, plum etc. can be grown till these do not interfere with the main mango crop. Vegetable crops as inter crops give better returns.

## Economics of mango based agroforestry system

The economics of mango based agroforestry system gave higher gross income, net return, B:C ratio than open farming system (sole crops and mango tree alone). The average cost of inter cropping would be Rs. 25,000 per ha and it would yield on an average of 15 tons per ha.

**Fig. 11.1:** Mango based agroforestry system (BAU, Ranchi, Jharkhand)

**Table 11.4:** B:C ratio analysis of different mango based agroforestry system in India

| Agro-forestry system | B/C ratio | References |
|---|---|---|
| Mango+Cluster bean | 1.40 | Das, A. K. 2016 |
| Mango+Cowpea | 2.01 | |
| Mango+Brinjal | 2.26 | |
| Mango+Okra | 2.21 | |
| Eucalyptus + Mango + Arhar | 1.18 | Annual Report (2017–2018), BCKV, West Bengal |
| Eucalyptus + Mango + Cowpea – Toria | 1.52 | |
| Eucalyptus+ Mango + Greengram – Toria | 1.47 | |
| Sole Arhar | 0.99 | |

| Agro-forestry system | B/C ratio | References |
|---|---|---|
| Sole Cowpea – Toria | 1.44 | |
| Sole Greengram – Toria | 1.35 | |
| Mango+Wheat (GW–273) | 2.53 | Ghode, B. D. and |
| Wheat (GW–273) | 2.37 | Dongre, R. 2017 |
| Mango+Mustard(Pusa bold) | 4.0 | Ghode, B. D. and |
| Mango+Mustard(Urvashi) | 3.62 | Dongre, R. 2017 |
| Mango+Mustard(NRCDR–2) | 3.78 | |
| Mango+Mustard(Arpan) | 5.08 | |
| Mango+Mustard(JM–3) | 4.86 | |
| Mango+ Cluster bean | 1.41 | Mishra, P. J. 2017 |
| Mango+ Cowpea | 2.10 | |
| Mango+ Brinjal | 2.38 | |
| Mango+ Okra | 2.30 | |
| Mango+ Blackgram | 1.33 | |
| Mango+Groundnut | 1.67 | |
| Mango+ Arhar | 2.36 | |
| Mango+ Sesamum | 1.37 | |
| Mango+Arhar | 2.48 | Mohapatra, S. C. and |
| Mango+ Groundnut | 1.19 | Behera, S. 2017 |
| Mango+ Blackgram | 1.24 | |
| Mango+ Cowpea | 1.32 | |
| Mango+ Black Gram | 1.35 | Mohapatra, S. C. and |
| Mango+ Cowpea | 2.27 | Behera, S. 2017 |
| Mango+ Brinjal | 2.40 | |
| Mango+ Okra | 2.31 | |
| Mango+Ginger | 1.87 | Gupta, M. 2018 |
| Mango+Turmeric | 2.29 | |
| Mango+Stevia | 2.30 | |
| Mango+Kalmegh | 3.00 | |
| Mango+Pea | 2.83 | Kumar, A. 2018 |
| Mango+Potato | 1.48 | |
| Mango+Garlic | 1.47 | |
| Mango+Tomato | 1.65 | |
| Mango+Watermelon | 2.62 | |
| Sole Mango | 0.43 | |
| Mango+Arhar | 2.53 | Prajnya, M. 2018 |
| Mango+Groundnut | 1.21 | |
| Mango+Blackgram | 1.36 | |
| Mango+Cowpea | 1.45 | |

| Agro-forestry system | B/C ratio | References |
|---|---|---|
| Mango+Mustard | 3.61 | Sahu *et al.*, 2019 |
| Mango+Linseed | 3.12 | |
| Sole Mango | 1.86 | |
| Sole Mustard | 1.76 | |
| Sole Linseed | 1.28 | |

## Conclusion

Crop productivity in many areas is low due to poor management of lands and weather aberrations. Monocropping neither provides gainful employment opportunity nor generates enough income to meet the family requirements. Fruit based agroforestry is considered as better alternative. Mango, being a versatile fruit crop, forms a potential component in many agroforestry systems. Mango based agroforestry systems are proven land use systems both economically and ecologically.

## References

Annual Report 2017–18 Yield and economics of *Eucalyptus tereticornis* + Mango based agroforestry system. Regional Research Station (Red & Laterite Zone), Bidhan Chandra Krishi Viswavidyalaya, Jhargram, West Bengal.

Das AK 2016 Evaluation of fruit based agroforestry systems in Nayagarh district of Odisha. M.Sc. thesis, Orissa University of agriculture and Technology, Bhubaneswar, Odisha.

Ghode BD and Dongre R 2018 Economics of Agroforestry system (Agrohorticulture system). Paper presented in annual group meeting of AICRP–Agroforestry, 28–30 July 2018, Birsa Agricultural University, Ranchi, Jharkhand.

Ghode BD and Dongre R 2018 Growth Characters of Mango, Mustard yield and Economics as influenced by different mustard varieties under agrihorticulture system. Paper presented in annual group meeting of AICRP–Agroforestry, 28–30 July 2018, Birsa Agricultural University, Ranchi, Jharkhand.

Gupta M 2018 Component interactions in mango based agroforestry system in the sub-tropics of Jammu and Kashmir. Ph.D. thesis, Division of Agroforestry, Faculty of Agriculture, Sher-e-Kashmir University of Agricultural Sciences and Technology of Jammu, Jammu-180009 (J & K).

Kumar A 2018 Studies on mango based agro-forestry system on farmer's field at Senha block, Lohardaga district, Jharkhand. M.Sc. thesis, Birsa Agricultural University, Ranchi, Jharkhand.

Mishra PJ 2018 Economics of mango based agroforestry system. Paper presented in annual group meeting of AICRP-Agroforestry, 28–30 July 2018, Birsa Agricultural University, Ranchi, Jharkhand.

Mohapatra SC and Behera S 2018 Economics of the agrihortisilvicultural system during 2017. Paper presented in annual group meeting of AICRP- Agroforestry, 28–30 July 2018, Birsa Agricultural University, Ranchi, Jharkhand.

Mohapatra SC and Behera S 2018 Performance of Mango and Cashew based Agroforestry systems under TSP during 2017. Paper presented in annual group meeting of AICRP– Agroforestry, 28–30 July 2018, Birsa Agricultural University, Ranchi, Jharkhand.

Prajnya M 2018 Evaluation of fruit based agrisilvihorticultural system with legume intercrops in rainfed uplands. M.Sc. thesis, Orissa University of agriculture and Technology, Bhubaneswar, Odisha.

Sahu S, Sah RB, Agrawal YA, Dhanyashri PV and Siwaji V 2019 Economics of mustard and linseed cultivation under mango based agri-horticulture system. *International Archive of applied sciences and technology* **10**(4) 85–88.

# CHAPTER - 12

# Coconut based Agroforestry System

## COCONUT

Coconut belongs to the palm tree family Arecaceae, which is an important mocotyledon plant and the only known living species of the genus Cocos. Its botanical name is *Cocos nucifera* L. Coconut is the single species of the genus Cocos belonging to the sub-family Cocoideae which includes 27 genera and 600 species. Coconut refers to the whole coconut palm, the seed, or the fruit, which botanically is a drupe, not a nut. The chromosome number (n) is 16 and 2n = 32. The term "coconut" is originated from the 16$^{th}$ century Portuguese and Spanish word coco, meaning 'head' or 'skull'. In India, the coconut palm is referred to as 'Kalpavriksha'– the 'tree of heaven' as each and every part of the palm is useful in one way or the other. Coconut plays a very important role in the Indian economy as ten million people in India directly or indirectly depend on coconut for their livelihood. Apart from the importance of copra and coconut oil which is broadly used in the manufacture of soaps, hair oil, cosmetics and other industrial products, its husk is a source of fibre which supports a sizable coir industry.

India ranks third in area and first in production of coconut in the world. As per the latest statistics available (2016–17), the annual coconut production in India is 23.90 billion nuts from an area of 2.08 million ha with an average productivity of 11481 nuts/ha. The four southern states of India *viz.* Kerala, Tamil Nadu, Karnataka and Andhra Pradesh are the major coconut producing states in India, which altogether contribute more than 90% of the area and production. The wet meat or kernel is rich in fat, carbohydrate, protein, fibre and moderate in its mineral content. Coconut water (endosperm) is rich in minerals and has a caloric value of 17.4 per 100 gm of water. It is also a rich source of fats, carbohydrates and minerals like phosphorus and calcium.

### Botanical Description

The coconut has adventitious root system continually produced from the basal 40 cm or so of the trunk, which is the swollen part or what is termed 'bole', in tall types and in some dwarf hybrids. It has no taproot and root hairs but has lots of primary roots which bear huge quantities of rootlets. The main root grows out somewhat horizontally from the bole and is mostly found within the topsoil. The stem develops from the single terminal bud called the 'cabbage' which is the palm's only vegetative growing tip. Under favourable condition, the establishment of the trunk of a young palm reaches full growth within 3–4 years. Leaves of a coconut seedling have the pinnae fused together and appear as entire leaves. After 8–10 leaves have been formed, subsequent leaves tend to split into leaflets. Inflorescence is enclosed in a double sheath or spathe, the complete structure known as a 'spadix' which is borne singly on the axil of each leaf. The palm is monoecious, *i.e.* its inflorescence contains both male and female flowers are present. Type of fruit is drupe, having three regions, is one seeded and outside is the husk.

## Climatic Requirements

Coconut is essentially a tropical plant growing mostly between 20° N and 20° S latitudes, but has been found to grow under varying agro-climatic regions. The average annual temperature for better growth and maximum yield is stated to be 27 °C with a diurnal variation of 6–7 °C and relative humidity more than 60 percent. The coconut palm is found at altitude up to 600 m above MSL. The coconut palm thrives well under an evenly distributed annual rainfall ranging from 127–250 cm. However, a well distributed rainfall of about 200 cm is the ideal rainfall for proper growth and higher yield.

## Soil

The coconut palm can grow on wide range of soil conditions. A number of factors such as drainage, topography, soil profile, soil depth, soil fertility and layout of the land have huge influence on the growth of the palm. Coconut palm prefers a well drained light textured soil permitting free root development and aeration. The major soil types suitable for coconut cultivation in India are laterite, alluvial, red sandy loam, coastal sandy, and reclaimed soils with pH ranging from 5.2–8.0.

Soil with a minimum depth of 1.2 m and good water holding capacity is preferred for coconut cultivation. Proper moisture either through well distributed rainfall or irrigation and sufficient drainage are essential for coconut plantation.

## Propagation/Planting material

Propagation is generally done by seed. In seed propagation, first of all selection of mother plant is done from which nuts have to be collected. It has great bearing on the productive life of palm. Vigorous seedlings which are one year old, having minimum of six leaves and girth of 10 cm at the collar level should be chosen for planting in the main farmland. Before time splitting of leaves in the seedlings could be a criterion for selecting good seedlings. However, 18–24 month old seedlings are preferred for planting in water logged areas.

## Planting

Size of the pit depends on the soil type and moisture contain in soil. In loamy soils with low water table, planting in pits of 1m × 1m × 1m filled with top soil to height of 50 cm is generally recommended. In laterite soils larger pits of the size 1.2m ×

1.2m × 1.2m should be dug and filled up with loose soil, farm yard manure, vermicompost, powdered cow dung and ash up to a depth of 60 cm before planting. While filling the pits, two layers of coconut husk can be arranged at the bottom of the pit with curved surface facing upwards for moisture conservation in soil. After arranging each layer, BHC 10% DP should be sprinkled on the husk to prevent termite attack and other insect. Laterite soils are known to be acidic in nature so common salt @ 2 kg per pit may be applied in six months interval and the floor of the pit should be dug to soften or break the hard pans.

### Spacing

In general square system of planting with a spacing of 7.5m × 7.5m (177 palms/ha) to 8.0m × 8.0m (156 palms/ha) is recommended for coconut. However, spacing of 7.5–10m is practiced in various coconut growing regions of India. If the triangular system is adopted, an additional 25 palms can be planted. Hedge system can also be adopted giving a spacing of 6.5m along the rows and 9.5m between rows. For multiple cropping in coconut gardens, it is advisable to go for wider spacing of 10m × 10m so as to provide ample opportunity to accommodate a number of perennial and annual crops in the interspaces. Which is also suitable for coconut based agroforestry system.

### Time of planting

Planting the coconut seedlings during month of May–June with the onset of pre-monsoon rain is perfect. In well drained soils, seedlings can be transplanted with the onset of south–west monsoon during June or with the onset of north–east monsoon during October–November.

**Table 12.1:** Important Varieties of Coconut

| Variety | Nut yield (nuts/palm /year) | Copra (yield kg/ palm/year) | Oil yield (kg/palm/ year) | Area for which recommended |
|---|---|---|---|---|
| | | Tall varieties | | |
| Chandra Kalpa | 100 | 17. 60 | 12. 67 | Kerala,Karnataka, Tamil Nadu, Andhra Pradesh, Maharashtra |
| Kera Chandra | 110 | 20. 79 | 13. 72 | Kerala, Karnataka, Konkan region (Maharashtra), Andhra Pradesh, West Bengal |
| Kalpatharu | 117 | 20. 50 | 13. 96 | Kerala, Karnataka, Tamil Nadu |
| Kalpa Pratibha | 91 | 23. 25 | 15. 50 | Kerala, Karnataka, Andhra Pradesh, Tamil Nadu, Maharashtra |

| Variety | Nut yield (nuts/palm /year) | Copra (yield kg/ palm/year) | Oil yield (kg/palm/ year) | Area for which recommended |
|---|---|---|---|---|
| Kalpa Dhenu | 86 | 20.81 | 13.63 | Kerala, Tamil Nadu, Andaman & Nicobar Islands |
| Kalpa Mitra | 80 | 19.25 | 12.80 | Kerala & West Bengal |
| Kalpa Haritha | 118 | 21.25 | 14.13 | Kerala, Karnataka |
| Kera Keralam | 109 | 19.18 | 13.05 | Kerala, Tamil Nadu, West Bengal |
| Kalpa Shatabdi | 105 | 28.65 | 18.34 | Kerala, Tamil Nadu, Karnataka |
| | | **Dwarf/Semi-tall varieties** | | |
| Chowghat Orange Dwarf | 112 | 14.34 | 9. 46 | All coconut growing areas |
| Kalpasree | 90 | 8. 64 | 5. 75 | Root (wilt) disease prevalent tracts |
| Kalparaksha | 87 | 11.94 | 7. 83 | Kerala & root (wilt) disease prevalent tracts |
| Kalpa Surya | 123 | 22.88 | 15.33 | Kerala, Karnataka, Tamil Nadu |
| Kalpa Jyothi | 114 | 16.19 | 9. 96 | Kerala, Karnataka, Assam |

**Table 12.2:** Hybrid varieties of Coconut

| Hybrid | Parentage | Nut yield (nuts/palm/ year) | Copra yield (kg/palm/ year) | Oil yield (kg/palm / year) | Area for which recommended |
|---|---|---|---|---|---|
| Chandra Sankara | COD × WCT | 110 | 25.00 | 16.82 | Kerala, Karnataka, Tamil Nadu |
| Kera Sankara | WCT × COD | 108 | 20.20 | 14.60 | Kerala, Karnataka, Maharashtra, Andhra Pradesh |
| Chandra Laksha | LCT × COD | 109 | 21.30 | 14.60 | Kerala, Karnataka |

| Hybrid | Parentage | Nut yield (nuts/palm/ year) | Copra yield (kg/palm/ year) | Oil yield (kg/palm / year) | Area for which recommended |
|---|---|---|---|---|---|
| Kalpa Samrudhi | MYD × WCT | 117 | 25.72 | 17.33 | Kerala, Assam |
| Kalpa Sankara | CGD × WCT | 84 | 14.62 | 9. 87 | Root (wilt) disease prevalent tracts |
| Kalpa Sreshta | MYD × TPT | 167 | 35.90 | 23.02 | Kerala, Karnataka |

## Manures and Fertilizers

Time to time manuring is necessary for better vegetative growth, early flowering, fruit bearing and sustainable yield of coconut palms. The first application of chemical fertilizers should be done after three months of planting. The optimum dose of fertilizer recommended for an adult palm is 0.5 kg N, 0.32 kg $P_2O_5$ and 1.2 kg $K_2O$/year. Green leaf/vermicompost/farm yard manure @ 50 kg per palm may be spread in the basin for better development.

**Table 12.3:** Nutrient Management in Coconut

| Age of coconut palm | May–June | | | September–October | | |
|---|---|---|---|---|---|---|
| | N | $P_2O_5$ | $K_2O$ | N | $P_2O_5$ | $K_2O$ |
| First year | Planting in May–June | | | 50 | 40 | 135 |
| Second year | 50 | 40 | 135 | 110 | 80 | 270 |
| Third year | 110 | 80 | 270 | 220 | 160 | 540 |
| Fourth year onwards | 170 | 120 | 400 | 330 | 200 | 800 |

In lateritic and acidic soils rock phosphate to be added as source of phosphorus is recommended. Fertilizers may be applied in two split doses during May–June and September–October for the rainfed palms and in four or more equal splits for irrigated palms avoiding the heavy rainfall period. In sandy soils with acidic pH, in addition to these fertilizers, 1 kg of dolomite may be broadcasted during April–May in the basins and incorporated into the soil by forking.

## Irrigation methods

Coconut responds well to summer irrigation *i.e.* irrigation water @ 40 litres per palm per week will increase the yield of nuts by 50 percent. Irrigation methods commonly adopted in coconut gardens are flooding, basin irrigation, sprinkler and drip irrigation. Under basin irrigation, 200 litres per palm once in four days will be beneficial. In dry areas where water is scarce drip irrigation system can be practiced. The amount of water recommended for drip irrigation in coconut is 66 percent of the open pan evaporation.

**Table 12.4:** Plant Protection of Coconut

| Major diseases | Damage | Control measures |
|---|---|---|
| Root wilt | Most important disease of coconut.Theearliestdiagnostic symptom is the abnormal inward bending of leaflets termed as ribbing or flaccidity, followed by rootdecay. | Complete eradication of all the disease affected palms, mixed cropping with cocoa and intercropping with tapioca, are some good measures. At initial stage of the disease, spray the crown and drench the basin with 500 ppm tetracycline. |
| Thanjavur wilt | Root decay. | Cut and remove all dead roots, cut ends of roots should be drenched with a mixture of 1.5g of aureofungin +1g copper sulphate. |
| **Major pest** | **Damage** | **Control measures** |
| Red palm weevil | Most destructive pest of coconut. Grubs cause damage inside the stem or crown by feeding on soft tissues. | Practice clean cultivation, avoid injury to trunk, attract weevils by smearing coconut toddy and destroythem. Injection of carbaryl (1%) to the infested trunk through a hole is alsouseful. |
| Rhinoceros beetles | Adult beetles usually cause damage, which bore into the tender fronds and spathes, unopened leavesand inflorescences. | Burn all dead trunks, leaves and logs. Spray carbaryl (0.1%) to kill the larvae. |

## Maturity

The tall varieties of coconut start flowering after 5 year and the dwarfs or hybrids variety of coconut flower from 3–4 year.

## Harvesting and Yield

Coconut palms mature in about 12 months after the opening of the spathe. It is the ripe coconut which is the main source of coconut products. Nuts which are 11 months old give good quality of fibre and can be harvested in the tracts where green husks are necessary for the manufacture of coir fibre. Economic rotation period of the coconut palm is about 60 years. Coconut palm can produce 12 inflorescences/year, but generally numbers of inflorescence per palm are less than 12. To get maximum yield of copra and coconut oil only fully mature nuts should be harvested.

After plantation crop management, the West Coast Tall yields on an average 80 nuts/palm/year in coastal Kerala, Karnataka etc. and the hybrids yield from 100–140 nuts/palm/year. The released hybrid varieties of coconut have a yield potential of 2.79–6.28 tonnes of copra/ha/year in comparison to 2 tonnes of copra yield realized by the tall cultivars which are predominantly cultivated by coconut farmers.

## Coconut in Agroforestry system

To maximise the utilisation of soil and sunlight in the coconut garden, intercropping can be adopted with a variety of crops like tubers and rhizomatous species (tapioca, elephant foot yam, sweet potato, greater yam, lesser yam, chinese potato, colocasia, ginger and turmeric), cereals and millets (paddy, sorghum, maize, pearl millet and finger millet), pulses and oilseeds (cowpea, green gram, black gram, red gram, ground nut, soybean, bengal gram and sunflower), vegetable crops (pumpkin, ash gourd, chillies, potato, french bean, snake gourd, amaranthus, brinjal, tomato, bottle gourd, ridge gourd, *coccinia* sp., dolichos bean, curry leaf and moringa), fruit crops (banana, papaya, pineapple and guava), flowering crops (*Heliconia sp., Anthurium sp.* and *Jasminum sp.)*, fodder grass(charabadam, stylo, berseem, napiar etc.) and legumes can be raised in coconut gardens upto 8–10 years.

During second growth phase of coconut (10–22 years of age) growing of other crops or shade loving crops in the interspace may be difficult due to poor sunlight availability. However, crops like colocasia, turmeric, ginger, some varieties of banana, fodder grass, shade loving medicinal plants etc. which can tolerate shade can be cultivated in this phase.

When coconuts attain a height of 5–6 m (above 20 years) *i.e.*, in older plantations, the crops mentioned in the initial stage and perennial plants cocoa, vanilla, black pepper, cinnamon, clove and nutmeg, sapota and medicinal & aromatic crops can be grown as mixed crops along with intercrops. By intercropping, raising fodder grasses such as hybrid napier or guinea grass along with leguminous fodder crops (charabadam, stylo, barseem etc.) in coconut garden has been found to be highly profitable which can support the rearing of milch animals.

**Fig. 12.1:** Coconut based agroforestry system (KAU, Kerala and USA, Banglore)

## Economics of coconut based agroforestry system

Coconut based agroforestry systems by raising compatible subsidiary crops and/ or integrating with livestock enables to increase the productivity, farm income and net returns from unit area of coconut plantations. The economics of coconut based agroforestry system gave higher gross income, net return, B:C ratio than open farming system (sole crops and cocnut tree alone).

**Table 12.5:** B:C ratio analysis of different Coconut based agroforestry system in India

| Agro-forestry system | B/C ratio | References |
|---|---|---|
| Coconut + *Alpina galangal* | 3.20 | Mohandas, S. 2011 |
| Coconut + *Aloe vera* | 3.80 | |
| Coconut + *Ocimum sanctum* | 3.00 | |
| Coconut + *Cymbopogan flexuosus* | 2.10 | |
| Coconut + *Pogestemon patchouli* | 2.10 | |
| Coconut + Guava + Aloe vera | 3.54 | Bari, M. S. and Rahim, M. A. 2012 |
| Coconut + Lemon + Aloe ver | 2.82 | |
| Open Aloe vera | 1.65 | |
| Coconut + Guava + Asparagus | 3.21 | |
| Coconut + Guava + Asparagus | 3.08 | |
| Open Asparagus | 1.87 | |
| Coconut + Guava + Misridana | 5.06 | |
| Coconut + Guava + Misridana | 3.83 | |
| Open Misridana | 2.33 | |
| Coconut + Heliconia (NPK 17:17:17) | 1.58 | Chowdappa, P. 2016 |
| Sole Heliconia (NPK 17:17:17) | 0.83 | |
| Coconut + Heliconia (NPK 13:05:13) | 2.73 | |
| Sole Heliconia (NPK 13:05:13) | 1.30 | |
| Coconut + HeliconiaVC(200g) + NC(100g) | 3.90 | |
| Sole HeliconiaVC(200g) + NC(100g) | 2.50 | |
| Coconut + Heliconia ½(VC + NC) + bio | 2.58 | |
| Sole Heliconia½(VC + NC) + bio | 1.20 | |
| Coconut + Heliconia ½(VC + NC) + 13:05:13NPK | 3.38 | |
| Sole Heliconia ½(VC + NC) + 13:05:13NPK | 1.90 | |

## Conclusion

In economic point of view, it was also pragmatic that coconut based agrgoforestry system with agriculture/horticulture/fodder/medicinal crops were found to be better yielding and more remunerative in comparison under multistoried agroforestry systems.

## Reference

Chowdappa P 2016 Floriculture for Enhancing Profitability of Coconut Gardens ICAR–CPCRI, Technical Bulletin No. 99.

Bari MS and Rahim MA 2012 Economic Evaluationand Yield Performanceof Some Medicinal Plants in Coconut Based Multistoried Agroforestry Systems. *The Agriculturists,* **10**(1) 71–80.

http://agritech.tnau.ac.in/horticulture/horti_pcrops_coconut_botany.html

Mohandas S 2011 Prospects of intercropping medicinal and aromatic plants in coconut garden. *Madras Agriculture Journal* **98**(1–3) 82–83.

CHAPTER - 13

# Cashewnut based Agroforestry System

## CASHEWNUT

Cashewnut (*Anacardium occidentale* L.) is a tropical evergreen tree which belongs to family Anacardiace, which includes many economically important tropical and subtropical trees. Cashew trees are native of Central and South America, very likely the centre of origin is Brazil. However large production is mainly confined to India, Mozambique, Tanzania, Kenya and Brazil. Though, cashew (*Anacardium occidentale* L.), is an exotic horticultural crop brought to India by Portuguese travellers in 16th Century but now adapted well in Indian conditions. It is grown along the coastal regions of Maharashtra, Goa, Karnataka and Kerala in the West Coast and Tamil Nadu, Andhra Pradesh, Odisha and West Bengal in the East Coast. It is extending in new areas such as Bastar region of Chhattisgarh and Plain regions of Karnataka, Gujarat, Jharkhand and in North Eastern Hill region. Major cashew producing states in India are Kerala, Karnataka, T.N., Andhra Pradesh, Goa, Maharashtra, Odisha, W.B., Chhattisgarh, Gujarat, Jharkhand and North Hilly States. During 2016–17, the production was 779.335 lakh tons of raw cashew from 10.41 lakh hactares area in India.

Cashew trees consist of about 73 genera and 600 species (Nakosone and Paull, 1998). Cashew is the most adaptable of all nuts. It is very important crop earning foreign exchange and having large employment potential. Cashew nuts are rich source of fats, proteins, carbohydrates, vitamin and other mineral like phosphorus and calcium.

### Botanical Description

The cashew tree is large and evergreen, fast growing and reaches a height of 10–15 with a short, often irregularly shaped trunk. The leaves are spirally arranged, leathery textured, elliptic to obovate with smooth margins. The flowers are produced in a panicle or corymb up to 26 cm long each flower is small, pale green at first, then turning reddish with five slender, acute petals. The largest cashew tree in the world covers an area around 7,500 $m^2$, it is located in Natal, Brazil.

The fruit of the cashew tree is an accessory fruit (called a pseudocarp or false fruit). Fruit is an oval or pear–shaped structure, a hypocarpium that develops from the pedicel and the receptacle of the cashew flower. Cashew apple ripens into a yellow or red structure about 5–11 cm long. It is edible and has a strong "sweet" smell and taste. The true fruit of the cashew tree is a kidney or boxing glove shaped drupe that grows at the end of the cashew apple. The drupe develops first on the tree and then the pedicel expands to become the cashew apple. The true fruit contains a single seed, which is frequently considered a nut in the culinary sense.

## Climatic Requirements

Cashew is essentially a tropical crop and grows best in the warm, moist and typically tropical climate. Most favorable temperature lies between 24 °C to 28 °C. Plain and hill slopcs upto 600–700 feet elevation are suitable. Cashew cannot tolerate frost and extreme cold for a long time.

## Soil

Cashew can be grown successfully in almost all soil types from sandy to lateritic soils but red sandy loam is best suited. Cashew is very sensitive to depth, stoniness and fertility of soil and availability of water in the area is also very important, but it cannot with stand water logging and excessive salinity or alkalinity. In India, cashew is mainly grown in laterite, red sandy loam and coastal sandy soil and also grown on black soils in Tamil Nadu and Andhra Pradesh to a limited extent.

**Table 13.1:** Hybrid Varieties of Cashewnut

| State | Varieties | Cross | Average yield (kg/tree) |
|---|---|---|---|
| Andhra Pradesh | BPP–I | T.No. 1 × T.No. 273 | 10 |
| | BPP–2 | T.No. 1 × T.No. 273 | 16 |
| | BPP–8 | Tree No. 1 × Tree No. 39 | 14 |
| | BPP–3, BPP–4, BPP–5, BPP–6 | — | 10–11 |
| Tamil Nadu | VRI–1, VRI–2, VRI–3, | — | 7–10 |
| Odisha | Bhubaneswar–1, Jagannath(BH6), Balabhadra(BH 85) | — | 7.4–10 |
| West Bengal | Jhargram–1, Jhargram–2 | — | 8.5–13.5 |
| Maharashtra | Vengurla–3 | 3Ansur–1 × Vetore 56 | 14.5 |
| | Vengurla–4 | Midnapore Red × Vetore 56 | 17.5 |
| | Vengurla–5 | Ansur Early × Mysore Kotekar 1/61 | 16.6 |

| State | Varieties | Cross | Average yield (kg/tree) |
|---|---|---|---|
| | Vengurla-6 | Vetore 56 × Ansur -1 | 13.80 |
| | Vengurla-7 | Vengurla-3 × M 10/4 (VRI-1) | 18.5 |
| | Vengurla-1, 2, 8 | — | 15-20 |
| Goa | Goa-1(Balli-2) | — | 7.2 |
| | Goa-2 (Tiswadi-3) | — | 5.5 |
| Kerala | BLA 39-4 (Madakkathara-1), NDR2-1(Madakkathara-2), Sulabha, Anakkayam-1, | | 13-17 |
| | Anagha | T 20 × K -30-1 | 13.7 |
| | Akshaya | H-4-7 × K -30-1 | 11 |
| | Dharasree | T 30 × Brazil-18 | 15 |
| | Amrutha(H1597) | BLA 139-1 × H 3-13 | 18.4 |
| | Priyanka(H 1591) | BLA 139-1 × K 30-1 | 16.9 |
| | Dhana (H 1608) | ALGD-1 × K 30-1 | 17.5 |
| | Kanaka (H1598) | BLA 139-1 × H 3 - 13 | 19 |
| Karnataka | Ullal-1, 2, 3, 4, UN-50, NRCC Selection-1 & 2, Bhaskara, H-130, Chintamani-1& 2 | — | 9-16 |

## Plant Propagation

Cashew is a cross pollinated crop and the orchards raised from seedlings show difference with respect to yield, nut size, apple colour and several other characters of tree. Hence, vegetative propagation is accepted in cashew as vegetatively propagated clonal progenies are genetically same to the mother plant or superior plant and give relatively consistent yield and come to fruiting earlier than seedling. Among other methods of vegetative propagation, softwood grafting was found to be the best for vegetative propagation of cashew and production of planting material of high yielding varieties through softwood grafting method has been a success story in cashew and is also found to have tremendous impact in improving cashew productivity. Many Regional Cashew Nurseries are developed under both public and private sectors and the farmers are always advised to acquire planting materials from recognized nurseries of recommended varieties on the basis of agro-climatic zone. Farmers can also produce planting materials after taking necessary training from the Directorate of Cashew Research, Puttur, Karnataka.

Soft wood grafting can be done throughout the year. However, during dry months (January–May) grafted plants should be protected by providing partial shade.

## Planting

The land is prepared with the start of pre–monsoon showers by clearing bushes, digging of pits, terracing the base of the trees etc. The square system of planting can be adopted. The best time for planting is usually monsoon season when the moisture is air surcharged (June–August) both in the west coast and east coast. If irrigation facilities are available, planting can be done all through the year except winter months. Pits of $1x1x1m^3$ size should be dug at spacing of 7.5m × 7.5m or 8m × 8m to accommodate 156 to175 plants per hectare. However the yields are very low in initial stages of plant growth, so for high density planting, 4m × 4m, 5m × 5m and 6m × 4m spacing are normally recommended. Later the pits should be filled to 3/4 level with a mixture of top soil, 10 kg of compost and 200g of rock phosphate.

High density planting method in cashew has provided higher yield and enhanced economic returns per unit area in the initial years, but also helped more well–organized use of inputs. Further, high density planting reduces the weed growth by canopy coverage, reduces soil temperature thus increasing the soil moisture content particularly during peak summer season and provides mulching effect. In high density planting, closer spacing 6.5m × 4m (384 plants/ha), 5m × 5m (400 plants/ha) and 5m × 4m (500 plants/ha) has given double more yield than normal spacing. Cashew nut yield can be increased to two times up to 6 years and four times up to 12 years in high density planting system. High density planting system however needs appropriate dwarf, high yielding and compact cashew varieties so that overlapping of canopy time to time is kept at minimum and also helped in the best breeding strategy for increasing productivity of cashew per unit area. Limb pruning and diagonal thinning in orchard can also be evaluated for their effectiveness in high density planting systems.

## Nutrient Management

The major nutrient requirement of cashew plant demands more liberal application of NKP is needed. Nitrogen and P are very important nutrients during the pre–bearing stage, but at the bearing stage, K together with N is also important for production. Studies conducted at DCR, Puttur, Karnataka recommended that application of 500g N, 125g P and 125g K and 10kg poultry manure per tree per year under normal density planting system (200 plants/ha) and 250g N, 50g P and 50g K and 10kg poultry manure per plant per year under high density planting system (625 plants/ha) is found better in terms of higher nut yield for rainfed cashew. The recommended dose of fertilizers as per age is given in table 13.2. The best application of enhance the efficiency, fertilizers should be applied in a circular trench of 10 cm depth at a distance of 0.5m, 0.75m and 1.5m away from the centre of the trunk in the $1^{st}$, $2^{nd}$ and $3^{rd}$ year onwards, respectively and covered with soil. Foliar sprays of nutrients (Urea 2–4%, DAP 1%, Orthophosphoric acid, $ZnSO_4$ 4%

and Cu 0.3–0.6%) at the emergence of the flush, panicle initiation and fruit set stages make sure better fruit set and also enhance nut yield in cashew.

### Fertilizer schedule (g/plant/year)

**Table 13.2:** Nutrient Management in Cashewnut

| Years after planting | Urea | Rock phosphate | Murate of Potash |
|---|---|---|---|
| 1 | 330 | 125 | 40 |
| 2 | 660 | 250 | 80 |
| 3 | 990 | 375 | 120 |
| 4 | 1320 | 500 | 160 |
| 5 and onwards | 1650 | 625 | 200 |

### Irrigation

Cashew is generally grown on degraded lands and experiences severe moisture stress from January–May, which adversely impact on flowering and fruit set. In regulate the rainwater harvest and to make it obtainable to the cashew plant during critical period, in situ soil and water conservation methods are very important. Cashew requires irrigation during summer months particularly in seedling stage. Protective irrigation especially summer months during January to march at fortnightly intervals @ 200 litres/plant improves fruit set, fruit retention, thus increasing nut yield.

**Table 13.3:** Plant Protection of Cashewnut

| Major disease | Damage | Control measures |
|---|---|---|
| Powdery mildew | Prevalent during rainy season, generally appears when the weather becomes cloudy. | Dusting with 2% sulphur W.P. is effective |
| **Major pests** | **Damage** | **Control measures** |
| Tea mosquito bug | The grubs cause the damage, making tunnels in irregular directions. As a result of injury, trees show different degrees of foliar yellowing, leading to shedding of leaves, drying of twigs and gradual death of the tree | Spray endosulfanor monocrotophos (0.05%) or carbaryl (0.1%) or quinalphos (0.05%) before infestation. |
| Stem and root borer | Most serious pest, responsible for a damageofnearly 25–30%. Bothnymphs and adults suck sap from the leaves, young shoots, inflorescence, developing young nuts andapples. | Remove and burn all the affected trees to avoid further spread of the pest. Inject 10 ml of dichlorvos (0.1%) per hole and plaster them with mud. |

## Harvesting

The plant starts bearing from the 3rd or 4th year, but full bearing is attained at 10th year and continues for another 20 to 25 years. The flowering started in month of December and continues for about three months. The fruits ripen during March–May or early June. After ripening fruits are collected every day and the nuts are separated. The peak picking months are March–May. Good quality nuts are grey green, smooth and well filled. After picking, the nuts are separated from the apple and dried in the sun for 2–3 days to bring down the moisture content to 10–12%. Properly dried nuts are packed in alkathene bags which can be stored for 6 months.

## Yield

A fully developed cashew plant yields about 25–30 kg of nuts per year.

## Cashew nut in Agroforestry system

Throughout the initial age (upto 3–5 yrs), many intercrops can be taken up between two rows of cashew plants to get more returns as well as utilize the light and soil resources efficiently. Pineapple can be grown as an intercrop between two rows of cashew for the first seven years. The spacing to be maintained for cashew is 8 m × 8 m (156 plants/ha) or 7.5 m × 7.5 m (175 plants/ha) or 10 m × 5 m (200 plants/ha). Growing pineapple in trenches across the slope between two rows of cashew helps to conserve the soil moisture, erosion etc. and also increases the yield of cashew by 1.5 times in agroforestry systems as compared to sole cashew. Cashew with pineapple can be grown as a more profitable intercrop under irrigated as well as rainfed conditions in west coast region in India. Another important suitable intercrops are tapioca, cucurbits, colocasia and elephant foot yam and after 7 years shade loving plant like turmeric, ginger etc. can be planted as intercrops.

In agroforestry system, planting with a spacing of 10 m × 5 m will be beneficial for growing intercrops such as colocasia, turmeric, elephant foot yarn, beans, cucurbits etc. in initial years of cultivation. Cashew nut based agroforestry system and hortisilvipastural models were laid out. Intercrops including crops, vegetables, short duration fruit crops, flowers, medicinal and aromatic plants, fodder crops and root crops were planted among these crops. The model helps in generating 6Fs of sustainability– food, fodder, fuel, feed, fibre and fertiliser. This species has also got the ability to grow on poor soils and can be intercropped with food crops such as maize, cassava and groundnuts.

## Economics of cashewnut based agroforestry system

The cashewnut based agroforestry system gave higher gross income, net return, and benefit cost ratio than open farming system (sole crops and guava tree alone). These types of systems are also highly profitable for farmes.

**Fig. 13.1:** Cashewnut based agroforestry system (UAS, GKVK, Bangalore and Gujarat)

**Table 13.4:** B:C ratio analysis of different Cashewnut based agroforestry system in India

| Agro-forestry system | B/C ratio | References |
|---|---|---|
| Cashew + Cowpea | 1.61 | Singh, S. 2009 |
| Cashew + Brinjal | 0.94 | |
| Cashew + Chilli | 0.66 | |
| Cashew + Colocasia | 2.11 | |
| Cashew + Okra | 1.54 | |
| Cashew + Pumkin | 1.67 | |
| Sole Cheshow | 1.51 | |
| Cashew + Black Gram | 2.03 | Mohapatra, S. C. and Behera, S. 2017 |
| Cashew + Cowpea | 2.75 | |
| Cashew + Brinjal | 2.67 | |
| Cashew + Okra | 2.61 | |
| Sole + Black Gram | 1.54 | |
| Sole + Cowpea | 2.35 | |
| Sole + Brinjal | 2.22 | |
| Sole + Okra | 2.47 | |
| Cashew + Arhar | 1.74 | Prajnya, M. 2018 |
| Cashew + Groundnut | 1.06 | |
| Cashew + Blackgram | 1.05 | |
| Cashew + Cowpea | 1.19 | |

## Conclusion

On the basis of economics, we can suggest farmer for cultivation of cashewnut based agroforestry system which will be help in income generation because the market price very high and cultivation costal or poor soil. Cashewnuts are considered as cash crop as they have got a huge marketed potential for export.

## References

Mohapatra SC and Behera S 2018 Crop yield and economics of agrisilvicultural system. Paper presented in annual group meeting of AICRP– Agroforestry, 28–30 July 2018, Birsa Agricultural University, Ranchi, Jharkhand.

Prajnya M 2018 Evaluation of fruit based agrisilvihorticultural system with legume intercrops in rainfed uplands. M.Sc. thesis, Orissa University of agriculture and Technology, Bhubaneswar, Odisha.

Singh S 2009 Studies on cashew based intercropping system with vegetable crops. M.Sc. thesis, Orissa University of agriculture and Technology, Bhubaneswar, Odisha.

CHAPTER - 14

# Aonla based Agroforestry system

## AONLA

Aonla (Botanical name- *Emblica officinalis,* family- Euphorbiaceae) or Indian gooseberry is indigenous to Indian sub-continent. Other Indian names are Dhatri, Amlaka, Adiphala (Sanskrit) Amla, Amlika, Aonla (Hindi). India ranks first in the world in area (1300 ha) and production (88200 tons) of this crop. Aonla is mostly cultivated in the states of Uttar Pradesh, Gujarat, Rajasthan, Maharashtra etc. The fruit is a good source of vitamin C. The fruit is having medicinal value. It has acrid, cooling, diuretic and laxative properties. Dried fruits are useful in haemorrhages, diarrhea, dysentery, anaemia, jaundice, dyspepsia and cough. Aonla is used in the indigenous medicines (Aurvedic system) *viz.* trifla and chavanprash. Fruits are commonly used for making preserves (murabbas), pickles, candy, jelly and jam. Besides fruits, leaves, bark and even seeds are being used for various purposes.

### Botanical Description

Aonla tree is small to medium in size, reaching 10–20 m in height. The branchlets are not glabrous or finely pubescent, 10–20 cm long, usually deciduous. Stem is smooth, greenish grey to brown, exfoliating bark, which peels off in thin flakes. The leaves are simple, sub sessile and closely set along branchlets, light green in color resembling pinnate leaves. The flowers are greenish–yellow. The fruit is nearly spherical, light greenish yellow, quite smooth and hard on appearance, with six vertical stripes or furrows. Ripening in autumn, the berries are harvested by hand after climbing to upper branches bearing the fruits. The taste of aonla is sour, bitter and astringent, and it is quite fibrous.

### Distribution

Aonla (*Emblica officinalis*), originally a native of Southeastern Asia, thrives well under various agro-climatic and agro-edaphic situations. It is geographically distributed in Bangladesh, China, India, Malaysia, Pakistan, Sri Lanka, Cuba, Puerto Rico, USA (Hawai & Florida), Iran, Iraq, Bhutan, Thailand, Vietnam, Philippines, Trinidad, Panama and Japan. In India, the Aonla is cultivated in many states including Uttar Pradesh, Gujarat, Rajasthan, Maharashtra and Haryana. Now the cultivation of this plant is rapidly spreading in other states like Tamil Nadu, Himachal Pradesh, Punjab and Andhra Pradesh

## Climatic requirement

Aonla being a sub-tropical crop prefers dry sub-tropical climate. The tree is well adapted to dry regions and heavy frost during the winter season is not suitable for its cultivation. An annual rainfall of 630–800 mm is ideal for its growth. The mature plants can tolerate freezing temperature as well as a high temperature up to 46$^{O}$C but young plants up to the age of 3 years should be protected from hot wind during May–June and from frost during winter months.

## Soil

Well drained fertile loamy soil is ideal for aonla cultivation, although it can be successfully grown in light as well as heavy soils except very sandy one. Slightly acidic to saline soils (pH 6.5–9.5, EC 5 dsm$^{-1}$, ESP 30–40) are suitable for its cultivation. Heavy soils with high water table are not suitable for its cultivation. Purely sandy soil should be also avoided.

## Varieties

Banarasi, Chakaiya, Francis, NA–4 (Krishna), NA–5 (Kanchan), NA–6, NA–7 (Promising variety), NA–10, BSR–1 (Bhavanisagar), Goma Aishwarya (released by CIAH, Bikaner) and Laxmi–52 (yielding potential of 2–2.5 q/tree, after 10 years).

## Propagation

**Seed Propagation–** Aonla can be raised from seeds but it produces small sized fruits of inferior quality as well as such trees have prolonged juvenile period. Plants raised from seeds are mainly used as rootstock, Mature fruits should be collected during November–December. Fruits are dried in open and seeds are extracted by applying light pressure. On an average, there are six seeds per fruit. One kg seed can be obtained from one quintal of Desi aonla fruits. Average test weight of aonla seed varies from 30,000–50,000 per kg or 300–500 per 10 g of fresh seed weight. For highest percentage of germination and to obtain seedlings with best root system, seeds should be soaked in 500 ppm $GA_3$ for 24 hours.

## Vegetative propagation–

**Budding–** It has been found to be most practical. Shield budding and patch budding are the commercial methods of aonla propagation. One year old seedlings witha girth of about 1 cm should be shield budded in early July with healthy and plump buds from new growth. Shield budding gives a success of 70–80 per cent.

**Grafting–** It is also one of the methods for raising good propagules. Soft wood grafting has been attempted with 70 per cent success at the site of terminal shoot of the rootstock raised in situ. It would be advisable to raise seedling rootstocks in situ and graft them with superior types of scion particularly in dry areas where mortality of budded plants is usually high.

## Planting time

**Rainy season–** July–September; **Spring season–** Mid of January–March

## Planting

Planting of aonla is mainly done in monsoon season during July–August. Pits of 1–1.25 $m^3$ size are dug two months prior to planting, *i.e.* in May–June, and left for 15–20 days exposing to sunlight for controlling weeds and insects. Each pit should be filled with surface soil mixed with 15 kg farm yard manure and 0.5 kg of phosphorus before planting the grafted seedlings. In sodic soil, 5–8 kg gypsum along with 20 kg sand is filled in the pit, irrigation is provided immediately after this. Ber, guava and lemon are usually planted in the centre of each square of aonla plants. Hedgerow planting is also being tried keeping line–to–line distance of 8 m., while plant to plant distance is reduced to 4–5 m.

- **Important Note:** At least 3 varieties should be planted in a ratio of 2:2:1 for the purpose of pollination and maximum yield.

## Nutrition

100 g N, 50 g P and 100 g K should be given to one year old plants. Dose is increased on yearly basis upto tenth year and thereafter a constant dose of 1kg N, 0.50 kg P and 0.75 kg K per plant/year should be applied. The fertilizer should be given in two split doses, in September–October and April–May.

## Irrigation

Young plants require watering during summer months at 15 days interval till they are fully established. Watering of bearing plants is advised at bi–weekly interval during summer months. After the monsoon rains, during October - December about 25–30 litres of water per day per tree through drip irrigation should be given.

## Training and pruning

The plants are trained to modified central leader system. Leaving only 4-5 well shaped branches with wide angle at about 0.75 m from the ground level, other dead, diseased, week crisscrossing branches and suckers should be pruned off at the end of December month.

## Mulching and Intercropping

During summer, the crop should be mulched with paddy straw or wheat straw at the base of the tree up to 15–20 cm from the trunk. Inter crops like green gram, black gram, cow pea and horse gram can be grown up to 8 years.

## Plant Protection

**Major insect:** Bark Eating Caterpillar (*Inderbella tetronis*)– Injection of Endosulphon 0.05% or Monocrotophos 0.03% in holes and plugging with mud is effective in protecting the tree against bark eating caterpillar.

**Major disease:** Rust (*Ravenellia emblicae*)– Spraying of Indofil M–45 @ 0.3% twice, first in early September and second 15 days after first application controls the spread of rust.

**Table 14.1:** Cost of Aonla Cultivation

| Particulars | Expenditure (Rs.) | Particulars | Income (Rs.) |
|---|---|---|---|
| **Cost** | | **Production** | |
| Preparation of farmyard | 2550 | Fruit 5 ton (after 4 years @ Rs. 12/kg) | 60000 |
| Plant cutting | 1224 | | |
| Sowing labour | 1224 | | |
| Fertilizers and irrigation | 1530 | Fruit 20 ton(after 8 years @ Rs. 12/kg) | 240000 |
| Weeding | 612 | | |
| Digging labour | 2040 | | |
| Other expanses | 1020 | | |
| **Total cost** | **10200** | **Total production obtained** | **300000** |

**Net profit**– Total production obtained– Total cost
60000–10200= 49800(after 4 years)
240000–10200= 229800(after 8 years)

## Harvesting and Yield

Fully developed fruits are harvested during February when they become dull greenish yellow from light green. Delay in harvesting results in heavy dropping of fruits in case of some varieties. A budded/grafted tree starts bearing fruit after 3 years of planting, whereas a tree developed from seedling may take 6–8 years. Vegetatively propagated aonla plants attain full bearing within 10–12 years and may continue to bear for 60–75 years of age under well managed conditions. A mature tree of about 10 years will yield 50–70 kg of fruit. After 10 years, aonla tree may bear 2–3 q/tree and 15–20 tons/ha. The average weight of the fruit is 60–70 g and 1 kg contains about 15–20 fruits.

## Utilization

Aonla (*Emblica officinalis* Gaertn.) is an important fruit plant. The fruit is highly nutritious and has tremendous medicinal value. It contains 500–1500 mg vitamin C, 0.2 mg nicotinic acid, 81.2% moisture, 14% carbohydrates, 3.4% fiber, 1.2% iron, 0.5% protein, 0.1% minerals and 0.02% phosphorus. Fresh fruit is used in the preparation of products like pickle and preserves and serves as an important ingredient in the preparation of Chyawanprash and Triphala, the well known ayurvedic medicine. Being the richest source of vitamin C, it helps cure scurvy, problems of teeth, gums, eye and stomach. Besides this, it is also used as remedy for dysentery, jaundice, anemia, bronchitis, cough, indigestion, urinary problems, septic fever, dyspepsia, etc. Owing to its property to strengthen the hair roots and maintain the color and lustre of hairs; it is used in a wide variety of hair products. It is bestowed with spasmolytic, purgative, expectorant, hypoglycemic, hepatoprotective, hypolipidemic, antibacterial and astringent attributes and also is a powerful antioxidant.

## Aonla in agroforestry system

Aonla based agroforestry system has immense potential to utilize and conserve rainfed area for betterment of poor farmers. Aonla being a deep–rooted deciduous tree species has a wide range of adaptability to grow on any type of soil. It is considered as a highly tolerant and potential fruit species suitable for growing under salt-affected and wasteland/ravine lands. Crops, vegetables, flowers and a few medicinal /aromatic plants are well suited for intercropping in aonla orchards. Cultivation of agricultural/horticulture/pasture/medicinal crops with aonla provides an opportunity for maximum use of land surface and helps to distribute the risk due to adverse climate. The tree canopy/crown of aonla with sparse foliage allows filtered light and permits intercropping even after the trees are fully grown. Intercropping not only generates an extra income, but also helps check soil erosion, fertility, leaching through ground coverage and improves the soil physic–chemical properties.

During initial 3–4 years of planting, aonla orchards present an excellent opportunity for utilizing hedgerow interspaces. Cereals, pluses, vegetables like bottle gourd, okra, coriander, cauliflower, pea, and turmeric; flowers like gladiolus and marigold have been found well suited for intercropping in aonla based agroforestry system. After the plants attain 5 years of age, generally shade loving crop like ginger, turmeric, medicinal plant etc. should be grown.

**Case study–1.** Fruit based diversified cropping system for arid region developed by Central Institute for Arid Horticulture (CIAH)

1. Aonla (*Emlica officinalis*) Based Cropping System– The tree canopy of aonla allows filtered light and permit intercropping even after it has made full growth. Two models have been developed–
   - M1= Aonla–Ber–Brinjal–Mothbean–Fenugreek.
   - M2= Aonla–Prosopis–Suaeda–Mothbean–Mustard.

   Highest gross return (25662 Rs. ha$^{-1}$) was obtained under M2 model.
2. Centre Agroforestry Research Institute (CAFRI), Jhansi has been working on aonla (*Emlica officinalis*) based agroforestry systems.

CAFRI initiated a field trial during 1989 with 4 varieties of aonla *viz.* Chakaiya, Kanchan, Krishna and NA–7 as fruit trees, subabul (*Leucaena leucocephala*) as multipurpose tree and blackgram (*Vigna mungo*) as intercrop in rainfed areas. At the age of 6 years, on an average the fruit yield from a plant was up to 93 kg. Besides fruit yield of aonla, 256 kg grain (on an average) was obtained from blackgram every year.

**Fig. 14.1:** Aonla based agroforestry system (BAU, Ranchi, Jharkhand)

## Economics of aonla based agroforestry system

Aonla provides higher economic returns with little investments in plantation establishment and its management.

**Table 14.2:** B:C ratio analysis of different Aonla based agroforestry system in India

| Agro-forestry system | B/C ratio | References |
|---|---|---|
| Aonla + Mothbean–fenugreek | 0.96 | Awasthi *et al.*, 2009 |
| Aonla + Mothbean–chickpea | 2.00 | |
| Aonla + Mothbean–mustard | 1.47 | |

| Agro-forestry system | B/C ratio | References |
|---|---|---|
| Aonla + Mothbean–cumin | 1.85 | |
| Sole Mothbean–fenugreek | 0.94 | |
| Sole Mothbean–chickpea | 1.43 | |
| Sole Mothbean–mustard | 1.41 | |
| Sole Mothbean–cumin | 1.32 | |
| Sole Aonla | 3.00 | Das *et al.*, 2011 |
| Aonla + turmeric | 6.29 | |
| Aonla + ginger | 3.45 | |
| Aonla + arbi | 3.20 | |
| Sole Turmeirc | 5.86 | |
| Sole Ginger | 3.38 | |
| Sole Arbi | 1.60 | |
| Aonla + Wheat (GW–273) | 2.63 | Ghode, B. D. and |
| Sole Wheat (GW–273) | 2.35 | Dongre, R. 2018. |
| Sole Aonla | 3.27 | Das, D. K. 2018 |
| Aonla+ Turmeric varieties R. Sonia | 3.40 | |
| Aonla+ Turmeric varieties NDH–92 | 4.15 | |
| Aonla+ Turmeric varieties R. Sonali | 3.44 | |
| Sole Turmeric varieties R. Sonia | 2.36 | |
| Sole Turmeric varieties NDH–92 | 3.56 | |
| Sole Turmeric varieties R. Sonali | 2.71 | |
| (Aonla+Turmeric) Control (No NPK) | 4.84 | Parihar, A. K. S. |
| (Aonla+Turmeric) 100% RDF through | 5.26 | 2018 |
| inorganic fertilizers | | |
| (Aonla+Turmeric) 100% of RDN through | 4.44 | |
| FYM | 6.56 | |
| (Aonla+Turmeric) 75% RDF through | | |
| inorganic fertilizers+25% RDN through FYM | 6.01 | |
| (Aonla+Turmeric) 50% RDF through | | |
| inorganic fertilizers+50% RDN through FYM | 4.74 | |
| (Aonla+Turmeric) 25% RDF through | | |
| inorganic fertilizers+75% RDN through FYM | 6.46 | |
| (Aonla+Turmeric) 100% RDF through | | |
| inorganic fertilizers+25% RDN through FYM | 5.65 | |
| (Aonla+Turmeric) 100% RDF through | | |
| inorganic fertilizers+Azotobacter | | |

## Conclusion

Economic analyses of different aonla based agroforestry systems showed that higher returns were obtained when the crops and other were grown in association with aonla, rather than sole cropping. So we suggest farmers for cultivation of aonla based agroforestry systems and enhance farm income.

## References

Awasthi OP, Singh IS and More TA 2009 Performance of intercrops during establishment phase of aonla (*Emblica officinalis*) orchard. *Indian Journal of Agricultural Sciences* **79**(8) 587–91.

Das DK 2018 Evaluation of Aonla (*Emblica officinalis*) based Agrihorticultural System with turmeric varieties. Paper presented in annual group meeting of AICRP-Agroforestry, 28–30 July 2018, Birsa Agricultural University, Ranchi, Jharkhand.

Das DK, Chaturvedi OP, Jha RK and Kumar R 2011 Yield, soil health and economics of aonla (*Emblica officinalis* Gaertn.)- based agri-horticultural systems in eastern India. *Current Science* **101**(6) 786–790.

Ghode BD and Dongre R 2018 Economics of Agroforestry system (Agrohorticulture system). Paper presented in annual group meeting of AICRP-Agroforestry, 28–30 July 2018, Birsa Agricultural University, Ranchi, Jharkhand.

http://www.nbrienvis.nic.in/WriteReadData/CMS/Emblica%20officinalis.pdf

Parihar AKS 2018 Effect of integrated nutrient management on turmeric (*Curcuma longa* L.) under Agri-silvi-horticulture system. Paper presented in annual group meeting of AICRP-Agroforestry, 28–30 July 2018, Birsa Agricultural University, Ranchi, Jharkhand.

Saini R and Sihag RC 2020 Flowering biology and phenology of Aonla (*Emblica officinalis* Gaertn.) in the semi-arid environment of North-west India. *Int. J. Agric. Res.* **15** 1–8.

## CHAPTER - 15

# Guava based Agroforestry System

### GUAVA

Guava, also called Apple of Tropics, King of Tropical Fruits, Poor man's apple, is a common tropical fruit cultivated in many tropical and subtropical regions. Guava (Botanical name *Psidium guajava*) is a small tree in family Myrtaceae, native to Mexico, Central America and Northern South America where it occurs wild. It was introduced in India in the seventeen century by Portugese. The most common types of guava include apple guava, yellow–fruited cherry guava, strawberry guava, and red apple guava. Guava is mostly eaten as raw fruit (ripe or semi-ripe).

Guava (Amrud) is one of the most common fruits in India. It ranks 4th on the basis of area and production after mango, banana and citrus. It is a very hardy sub-tropical or tropical plant. Keeping in mind its multi–utility, nutrition, and health benefits people commonly call it as "poor man's fruit" or "apple of the tropics". Guava contains Vitamin C in high concentration. In India, area covered by guava is 1.30 lakh ha in Uttar Pradesh (largest area and production), Bihar, M.P, Maharastra and Andhra Pradesh. Prayagraj (UP) is very famous for growing the best guava in the country as well as in the world.

### Distribution

Globally, the major guava producing countries are South Asian countries (India, Sri–Lanka, Vietnam, Indonesia, Bangaladesh etc) and the Hawaii Islands, Cuba. In 2019, world production of guavas was 46.5 million tons, led by India with 41% (18.8 million tons) of the total production followed by China (10%) and Thailand (7%). In India, major guava producing states include Uttar Pradesh, Bihar, West Bengal, Maharashtra, Chhattisgarh, Tamil Nadu, Karnataka, Madhya Pradesh, Gujarat and Andhra Pradesh.

### Botanical Description

Guava is a fast growing evergreen shrub or small tree that can grow to a height of 3 to10 meter. It has a shallow root system. Guava produces low drooping branches from the base and suckers from the roots, the guava is easy to recognize because of its smooth, thin, copper-colored bark that flakes off. Young twigs are quadrangular and downy. The leaves, aromatic when crushed, are evergreen, opposite, oval, leathery, and with conspicuous parallel veins. Flowers are white in color, are faintly fragrant, borne singly or in small clusters on the leaf axils, the type of inflorescence is Solitary. It is self-pollinated crop. The fruit, exuding a strong, sweet, musky odor when ripe, type of fruit is Berry and the edible parts are thalamus and pericarp.

## Cultivation Technology

### Climate

Guava plants are hardy in nature, can be grown successfully in tropical and subtropical regions up to 1500 m above mean sea level (msl). Best quality guavas are found where minimum night temperature of 10 °C prevails during winter season. The overall temperature of 23–25 °C is good for growth and production. However, it tolerates high temperatures and drought conditions in Northern India in summer season (April–June) but it is susceptible to severe frost as it can kill the young plants. An annual rainfall 100 cm is enough during the rainy season (July–September) but rainfall during harvesting period deteriorates the quality of fruits.

### Soil

Guava is planted on varied types of soils in heavy clay to very light sandy soils. Yet, very good qualities of guavas are produced in river basins areas of India. It tolerates a soil pH of 4.5–8.2 and maximum concentration of its feeding roots is available up to 25 cm of soil depth. Thus the top soil should be fairly rich to provide sufficient nutrients for accelerating new growth which bears fruits. The crop is sensitive to water-logging so drainage necessary.

### Varieties

1. Lucknow 49/L–49– Also known as Sardar. It is bronzing resistant variety.
2. Allahabad Safeda– It can withstand drought condition. It is guava wilt resistant variety.
3. Chittidar– This variety is very popular in we stem Uttar Pradesh.
4. Harijha–Harijha is more popular in Bihar because of profuse bearing.
5. Hafshi– It is red fleshed guava having good taste.
6. Apple Colour– Its fruits are medium sized and pink coloured.
7. Saharanpur Seedless
8. Nagpur Seedless
9. PusaSrijan– It is good for high density planting.
10. ArkaMridula– This is a seedling selection of variety Allahabad Safeda. It is seed less and dwarfvariety.
11. Allahabad Surkha– Allahabad Surkha is an outstanding variety of large, uniform pink fruits
12. with deep pink flesh.
13. Lalit– It isa new variety of Guava had been released by the CISH, Lucknow, for commercialcultivation. It is also good for Rajasthan climate.

14. Pant Prabhat- It had been selected by the Department of Horticulture, GBPU&T, Pantnagar (Uttrakhand).
15. Behatcoconut- It is seedless variety.
16. Allahabad Round-Parthenocarpic variety.

**Table 15.1:** Hybrid Varieties of Guava

| Hybrid | Parentage | Importance characters |
|---|---|---|
| ArkaAmulya | ArkaSafeda × Seedless | — |
| Arka Rashmi | Kamsari × Purple Local | Pink flesh variety |
| Arka Kiran | Kamsari × Purple Local | Pulp is firm and deep pink in colour |
| KohirSafeda | Kohir × Allahabad Safeda | — |
| Safed Jam | Allahabad Safeda × Kohir | — |
| Hybrid–45 | Allahabad Safeda × L–49 | — |

## Propagation

Guava is propagated both by seeds and vegetatively. Guava is generally propagated through seeds. Seeds germinate in 2 to 3 weeks but may take as long as 8 weeks. Guava trees are vegetatively propagated by inarching or air layering and budding. It is commercial propagated by Stooling and Air Layering. It has been grown also from root cuttings, pruned branches and budding.

Guava plants that have been vegetatively propagated start bearing fruits 2 to 3 years after planting and are fully productive at 8 to 9 years. Guava trees propagated generally from seed require more time for fruit production. In India, 8 to 10 year old trees from seedlings may produce numbers of 400–500 fruits per year but grafted trees at the same age may produce 1000–2000 fruits.

## Planting Time

Planting is done during the rainy season. June and July is the ideal time for planting the vegetative propagation and seedling.

## Planting

Square system of planting is generally adopted. Pits of 1m × 1m × 1m size are dug before the monsoon and filled with a mixture of farmyard manure and soil. The plants are usually planted at a distance of 5–8 m. The planting distance is decided according to variety, soil fertility and availability of irrigation facilities and purpose of planting like agroforestry, forestry, farm boundary etc.

Standard spacing is 6m × 6m accommodating 278 plants/ha and by increasing the plant density, productivity can be increased. A high density planting cause erect growth of branches making the plant tall, compact and also gives higher yield per unit area in early years of fruiting.

## Manure and Fertilizer

The time application of fertilizer depends on the crop variety, area and region. In Northern India, fertilizer is applied in the first week of May for rainy season crop and in first week of October for winter season crop. The trees are manured twice a year, first during June to July and second in October.

A fertilizer dose of 600 g N and 400 g K in Northern Region, 260 g N, 320 g P and 260 g K in Eastern Region, 900 g N, 600 g P and 600 g K in Southern Region and 600 g N, 300 g P and 300 g K per plant per year in Western Region is recommended.

## Irrigation

Guava plants are mostly grown under rainfed condition. In winter season, irrigation is required at an interval of 20 to 25 days and in the summer season it is provided at an interval of 10 to 15 days by the ring method to young guava plants. Drip irrigation has also proved to be very beneficial for guava where about 60% of the water used for irrigation is saved. Substantial increase in size and number of fruits has been observed due to frequent irrigation.

## Training and pruning

Training of plants in juvenile stage is necessary in order to build a strong structure and to avoid weak crotches. The plants are trained as low headed trees to facilitate multiple hand pickings. Pruning is usually recommended after harvesting or in spring. Fruiting trees are pruned to check overcrowding in the guava orchard. Best time of pruning is Feb–March, Summer pruning is generally avoided as the plants get damaged due to sun burn.

## Harvesting and yield

Guava fruit matures in 90 to 150 days after flowering. It is ready for harvest as soon as the deep green colour turns light and a yellowish green patch appears. The plants start bearing fruit at an early age of 2–3 years but they attain full bearing capacity at the age of 8 to 10 years. The yield of a tree depends on its age, cropping pattern, area, climate and the cultural practices. After 10 years, tree yields about 100–150 kg of fruits per year. In agroforestry system different type of crops are taken in both rainy and winter season, more yields may be obtained in the rainy season.

Guava is harvested throughout the year (except during May–June) in one or the other area of the country and peak harvesting periods in north India are August month for rainy season crop, November and December for winter season crop and March to April for spring season crop. In the mild climatic conditions of the other parts of the India, the peak harvesting periods are not so distinct.

Guava fruits develop best flavour and aroma after ripening. In most of the commercial varieties of guava, fruit ripening is indicated by the colour development which is usually yellow or pale green. Fruits are harvested selectively by hand along with the stalk and leaves.

## Yield

Guava can be harvested round the year. Average yields of fruit from improved trees may be between 400–500 fruits/tree which comes around 40–50 kg/tree, 25–30 t/ha and up to 50 t/ha have been obtained.

## Storage

The fresh fruit has a short shelf life and distant marketing can be done only if it is properly stored and packaged. The shelf life can be extended upto 20 days by storing them at low temperature of 5 °C and 75–85% relative humidity. It can be stored for about 10 days at room temperature (18–23 °C) in polybags providing a ventilation of 0.25%.

## Utilization

The fruit is a good source of vitamin C (299 mg/100g), pectin, calcium and phosphorus. The fruit is used for the preparation of processed products like candies, preserves, jellies, jams and as a marmalade jam served on toast. Guava jelly puree is very accepted for its eye–catching purplish–red colour, pleasant taste and aroma. The guava puree can be used in juice, cakes, puddings, sauces, ice–cream, jam, jelly and many other products. Good quality salad can be prepared from the shell of ripe fruits. Leaves of guava are used for curing diarrhoea and also for dyeing and tanning. It is also used in wood, firewood, leaves and bark for tannin (10% in the leaves on a dry weight basis, 11–30% in the bark). Guava tree wood is useful for various tool manufacturing, fences or firewood (mainly good source of charcoal).

## Guava in Agroforestry system

Under guava based agroforestry system generally leguminous crops or vegetable can be grown as intercrops during the first three years of planting provided irrigation facility is available. This land use system is also capable of providing employment opportunities to farmers. Guava based agroforestry system generates additional employment and income for farm families and improves the socio-economic status of growers. Interspace between the rows of trees can be successfully utilized for cultivation of agricultural crops, fruits, vegetable, spices, medicinal and aromatic plants or fodder crops.

Guava, being a productive and valuable fruit bearing plant of dwarf stature canopy can be very suitable as second storey crop for additional income generation at least in the initial years of the guava plantation. Among the different combinations of crops, the maximum cumulative paddy equivalent yield of 51.24 ton/ha was recorded in case of Mango + Guava + French bean and among the different filler plants, guava was found to be the most promising tree species. Orwa *et al.*, 2009 reported that in agroforestry systems, guava can be intercropped with fodders plants such as maize, sorghum, cowpea and berseem for high income. Raturi *et al.*, 1993 reported that in Gujarat, guava trees have been planted in association with *Leuceana leucocephala* (subabool), *Cenchrus ciliaris* (buffel grass) and *Stylosanthes fruticosa* (African stylo) in agri-silvi-pastoral systems to increase productivity and reclaim eroded, degraded and compacted soils with gullies and rocky area Gujarat, India.

**Fig. 15.1:** Guava based agroforestry system (BAU, Ranchi, Jharkhand)

## Economics of guava based agroforestry system

The economic evaluation of guava based agroforestry system shows that it gives higher gross return, net return and benefit: cost ratio than open farming system (sole crops and guava tree alone). The agroforestry systems have given an additional income of 4629–6576 rupees per year. Fruit crop based agroforestry systems have given higher B:C ratio as compared to mono-cropping. Average compound income,

Net Product Value (NPV) and Internal Rate of Return (IRR) have been highest with arable crops/horticulture/medicinal/fodder crop etc.

**Table 15.2:** B:C ratio analysis of different Guava based agroforestry system in India

| Agro-forestry system | B/C ratio | References |
|---|---|---|
| Guava + Rice variety RCPL–1–24 | 2.01 | Bhatt, B. P. and Misra, L. K. 2003 |
| Guava + Rice variety RCPl–1–25 | 1.72 | |
| Guava + Rice variety RCPL–1–29 | 1.69 | |
| Sole Rice variety RCPL–1–24 | 1.44 | |
| Sole Rice variety RCPl–1–25 | 1.32 | |
| Sole Rice variety RCPL–1–29 | 1.22 | |
| Coconut+Guava+*Aloe vera* | 3.54 | Bari and Rahim 2012a |
| Coconut+Lemon+*Aloe vera* | 2.82 | |
| Sole *Aloe vera* | 1.65 | |
| Coconut+Guava+*Asparagus racemosus* | 3.21 | Bari and Rahim 2012b |
| Coconut+Lemon+*Asparagus racemosus* | 3.08 | |
| Sole *Asparagus racemosus* | 1.87 | |
| Coconut+Guava +*Kaempferia angustifolia* | 5.06 | |
| Coconut+ Lemon +*Kaempferia angustifolia* | 3.83 | |
| Sole *Kaempferia angustifolia* | 2.33 | |
| (Guava + Arrowroot) Control (No Fertilizer) | 1.17 | Biswal, S. S. 2018 |
| (Guava + Arrowroot) STD (50%) | 1.45 | |
| (Guava + Arrowroot) STD (50%) + FYM | 1.45 | |
| (Guava + Arrowroot) STD (50%) + FYM+BF | 1.60 | |
| (Guava + Arrowroot) STD (75%) | 1.63 | |
| (Guava + Arrowroot) STD (75%) + FYM | 1.74 | |
| (Guava + Arrowroot) STD (75%) + FYM+BF | 1.84 | |
| (Guava + Arrowroot) STD (100%) | 1.79 | |
| (Guava + Arrowroot) STD (100%) + FYM | 1.88 | |
| (Guava + Arrowroot) STD (100%)+ FYM+BF | 1.97 | |
| Guava + Pearl millet | 2.00 | Gupta, S. 2019 |
| Sole Pearl millet | 1.79 | |
| Guava 75% RDF through water soluble fertilizers | 3.65 | Janaki *et al.*, 2020 |
| Guava 100% RDF through water soluble fertilizers | 3.03 | |
| Guava 75% RDF through basal application of conventional fertilizers | 2.65 | |
| Guava 100% RDF through basal application of conventional fertilizers | 2.51 | |
| Guava Control | 2.42 | |

## Conclusion

Guava based agroforestry is economically viable and more profitable than many of crop rotations followed in the study area. On the basis of economics guava based agroforestry system can be suggested to farmers which will be helpful in higher income generation. Fruit crop based agroforestry system is a multi-pronged programme of land use system to meet various requirements and could be ideal for restoration of ecological balance, improvement in farm profits and minimization of soil fertility depletion and loss of soil through erosion.

## References


Bari MS and Rahim MA 2012a Bio-economic evaluation of *Aloe vera* in coconut based multistoried agroforestry systems. *J. Agrofor. Environ.* **3**(2) 61–63.

Bari MS and Rahim MA 2012b Economic evaluation and yield performance of some medicinal plants in coconut based multistoried agroforestry systems. *The Agriculturists* **10**(1) 71–80.

Bhatt BP and Misra LK 2003 Production Potential and Cost-Benefit Analysis of Agrihorticulture Agroforestry Systems in Northeast India. *Journal of Sustainable Agriculture* **22**(2) 99–108. DOI: 10.1300/J064v22n02_07

Biswal SS 2018 Nutrient balance studies of guava + arrowroot agroforestry system. M.Sc. thesis, Orissa University of Agriculture and Technology, Bhubaneswar, Odisha.

Gupta S 2019 Effect of NPK and lime levels on pearl millet (*Pennisetum glaucum*) under guava (*Psidium guajava*) based agri-horti system. M.Sc. thesis, Institute of Agricultural Sciences, Banaras Hindu University, Varanasi, U.P.

Janaki D, Poorniammal R and Rajangam J 2020 Effect of drip fertigation on yield and quality of guava under ultra high density planting in Tamil Nadu. *International Journal of Chemical Studies* **8**(2) 2720–2722.

Orwa C, Mutua A, Kindt R, Jamnadass R and Anthony S 2009 Agroforestry Database: a tree reference and selection guide version 4.0. World Agroforestry Centre, Kenya.

Raturi GB and Hiwale SS 1993 Horti-silvi-pastoral system for increased productivity of marginal and degraded lands under rainfed conditions. *Adv. Hort. Forestry* **3** 179–186.

CHAPTER - 16

# Medicinal and Aromatic Plants based Agroforestry system

The plant products obtained from different parts like roots, stem, bark, leaves, twigs, flowers, fruit and seeds either individually or collectively or chemicals derived from these parts are used in different systems of medicine especially for their curative properties are known as Medicinal plants. Plants that contain aromatic essential oils that are extracted for perfumery, cosmetics, flavoring, medicinal and other human uses are otherwise called as Aromatic crops.

Medicinal Plants or Medicinal herbs are essence of very traditional medicine practice. The treatment using herbs developed in different parts of world with different names. Herbalism is the systematic study of medicinal herbs and its botany which is mainly focusing its medicinal uses. Ethno–medicine is the study of traditional medicine use or practiced.

Medicinal and aromatic plants (MAPs) play a very important role in the healthcare of people and others around the world, particularly in developing countries. MAPs are important from several perspectives like preventive and curative treatments and are source of income to underprivileged or aboriginal people who collect these herbal plants from forests or cultivate on their field.

## Area and Production

More than 8,000 medicinal plants and about 1500 aromatic plants reported in India. Rajasthan is having highest area under these crops with a share of 56% followed by Uttar Pradesh (25%) and Madhya Pradesh (11%). In case of production, Madhya Pradesh ranks first with 44% share, Rajasthan with 19% and Tamil Nadu 16%. Mint (*Mentha arvensis*) and Aonla (*Emblica officinalis*) covered largest areas (>1.0 lakh hectares each) followed by Isabgol (*Plantago ovata*), Senna (*Cassia angustifolia*), Ashwagandha (*Withania somnifera*) and other crops. Cultivation of medicinal and aromatic plants has increased from 262000 ha during 2005–06 to 633900 ha in 2015–16. Similarly, the production has increased from 202000 tons in 2005–06 to 1022500 tons in 2015–16 with an annual growth rate of 2.76 per cent per annum.

## Trade and Market

The ranks of India second in the world after China in export of medicinal plants. The trade of herbal medicines is about Rs. 27 billion per year in global market. A medicinal plant contributes 15–30% to the total income of poorer households. About 25% of drugs in modern pharmacopoeia are derived from plants. The importance of medicinal plants is continuous increasingly being recognized from ecological, social and economic perspectives. The reliance on medicinal plants as primary source of healthcare and meeting the growing demand of industries is creating heavy pressure on medicinal plant populations due to over harvesting.

National Medicinal Plants Board (NMPB) is exploring many ways to augment their cultivation. Right now the market size is of 4.2 billion rupees and it is estimated that it will be of 14 billion rupees by 2026. So we have to be ready for meeting the demand. Several types of raw materials are herbs and shrubs, which can be grown and harvested in a period of one year. But we have to improve cultivation of massive medicinal trees. About 80% of the raw materials are collected from the forests. Sustainable livelihood management of the whole supply chain from the tribal people who collect them from the forests to the traders, processors, packagers and the industry is the key to ensure enough supply of raw materials. Remaining 20–25% is cultivated by farmers for which sustainable cultivation and harvesting should be promoted. Farmers should be ensured enough income to encourage them to cultivate medicinal plants. The NMPB is promoting integrated cultivation of medicinal and aromatic plants.

## Medicinal and Aromatic Plants in Agroforestry system

Medicinal plants based agroforestry systems can play an important role through which tree crops as well as medicinal plants can be cultivated on same piece of land and gives an additional income to farmers. Since it allows combined cultivation of trees and medicinal plants/annuals as well, this type of system can also reduce pressure on natural forests. A lot of the MAPs are grown under forest cover and are shade tolerant therefore agroforestry offers a suitable strategy for promoting their cultivation and conservation. Many approaches are feasible for this, *e.g.* integrating shade tolerant MAPs as lower strata species in multi–strata systems; cultivating short rotation MAPs as intercrops in existing stands of plantation tree–crops (agrisilviculture) and new forest plantations, growing medicinal trees as shade providers, boundary markers, control soil erosion and on soil conservation structures interplanting MAPs with arable crops involving them in agroforestry programs.

Agroforestry is the best option for commercial cultivation of MAPs in India and exploratory studies on incorporation of various MAPs under different timber, fruit, flower, fuel and fodder based agroforestry systems have been carried out and advocated being ecologically, socially and economically sound. The area under agroforestry has shown a remarkable increase which offers a convenient strategy for promoting cultivation and conservation of MAPs. In addition to available area under major tree crops, the area under plantation crops such as coconut, cashewnut, arecanut and cocoa amounting to 2140, 1011, 452 and 41 thousand hectares, respectively, hold great potential for MAPs cultivation. The research on integration of various MAPs under these plantation crops has been accepted and outcomes are ecologically and economically positive under such tree and crop combinations. The studies have revealed that the wider spatial arrangement of trees provide better alley spaces for cultivation of MAPs particularly shade loving species.

Big trees and perennial medicinal trees that need to be cultivated at wider spacing such as African cherry (*Prunus africana*), nilgiri (*Eucalyptus globule*) for oil, sandalwood (*Santalum album*), ashok (*Saraca indica*), bael (*Aegle marmelos*), custard apple (*Anona squamosa*), aonla (*Emblica officinalis*), harre (*Terminalia chebula*), bahera (*Terminalia bellirica*) drumstick (*Moringa oleifera*) and soapnut tree (*Sapindus mukorossi*) can be intercropped with annual crops or other in the early years until the tree canopy covers the ground. Several of the medicinal trees may allow intercropping for many years or on a fixed basis depending on the spacing and nature of the trees. The time to time intercrops give some income to farmers during the period when the main trees have not started production.

Many well known medicinal plants that have been successfully intercropped with fuel wood trees (*e.g.*, *Acacia auriculiformis, Acacia nilotica, Prosophic* spp., *Albizia lebbeck, Bombex ceba Eucalyptus tereticornis, Gmelina arborea* and *Leucaeana leucocephala*) in India, include safed musli (*Chlorophytum borivilianum*), sarpgandha (*Rauvolfia serpentina*), ashwagandha (*Withania somnifera*) turmeric (*Curcuma longa*), wild turmeric (*C. aromatica*), kali musli (*Curculigo orchioides*) and ginger (*Zingiber officinale*).

Maheswari *et al.* (1985 and 1995) brought out that growing of rauvolfia and pachouli as intercrops can provide extra income to coconut growers. *Aconitum heterophyllum* (Atis), *Asparagus racemosus* (shatavari), *Amomum subulatum* (Large cardamom), *Amomum villosum* (Saren), *Chlorophytum borivilianum* (Safed musli), *Costus speciosus* (Crepe ginger), *Dioscorea deltoidea* (Himalayan yam), *Aloe vera* (Ghrit kumari), *Panax ginseng* (Ginseng) and *Rauvolfia serpentina* are few commercially highly valuable are suitable as understory component of agroforestry. Experiments at different sites in India established that aromatic plant like Mentha spp. (*M. arvensis, M. piperita, M. citrata var. citrata, M. spicata, M. cardiaca and M. gracilis*) and Cymbopogon spp. lemon grass (*C. flexuosus*); Java citronella (*C. winterianus*); and palmarosa (*C. martinii*) and *Ocimums pecies* can be intercropped with commercially important tree species.

Mohan *et al.* (2006) reported presence of many medicinal plants in all the 75 home gardens surveyed in Thrissur district in Kerala and were considered very important by the home garden owners with fruit trees as main crop. The research reported medicinal plants as 15–17% of the total economically important species having different curative properties, domestic use, abundant in all categories of home gardens.

**Fig. 16.1:** Medicinal and Aromatic Plants based agroforestry system (BAU, Ranchi, Jharkhand)

The medicinal plants are chosen to suit the topography of the site. Shade loving plants of family Araceae and Zingiberaceae can be selected for intercropping as per the choice of demand and market (Banik, 2000). Species like *Aloe vera, Catharanthus roseus, Cassia angustifolia, Curcuma domestica, Plantago ovata, Withania somnifera*, etc. can tolerate shade and are suitable to grow as intercrop due to their wide adaptability.

### Economics of Medicinal and Aromatic Plants based agroforestry system

Many medicinal plants have been evaluated and compared with the other crops or existing farming systems for economic feasibility. Many medicinal plants have been evaluated and compared with the other crops or existing farming systems for economic feasibility.

**Table 16.1:** B:C ratio analysis of different Medicinal and Aromatic Plants based agroforestry system in India

| Agro-forestry system | B/C ratio | References |
|---|---|---|
| Tamarind + Safed musali | 1.94 | Kumar *et al.*, |
| Sole Safed musali | 1.51 | 2010 |
| Peach + Grewia + Setaria + Mucuna | 2.32 | Verma and |
| Peach + Morus + Setaria + Mucuna | 2.35 | Thakur 2010 |
| Peach + Setaria + Mucuna | 2.41 | |
| Grewia + Setaria + Mucuna | 2.18 | |
| Morus + Setaria + Mucuna | 2.37 | |
| Sole Mucuna | 1.77 | |
| Coconut + *Andrographis paniculata* | 2.56 | Basavaraju |
| Coconut + *Solanum nigrum* | 2.68 | *et al.*, 2011 |
| Coconut + *Coleus forskolii* | 1.66 | |
| Coconut + *Ruta graveolens* | 2.79 | |
| Coconut + *Lepidium sativum* | 1.92 | |
| Coconut + *Ocimum sanctum* | 2.71 | |
| Coconut + *Maranta arundinacea* | 2.28 | |
| Coconut + *Kaemferia galangal* | 1.27 | |
| Coconut + *Mucuna pruriens* | 2.41 | |
| Coconut + *Hibiscus sabdariffa* | 1.81 | |
| Coconut + *Abelmoschus moschatus* | 1.58 | |
| Coconut + *Cymbopogon winteranus* | 1.96 | |
| Coconut + *Cymbopogon flexuosus* | 2.89 | |
| Coconut + *Vetiveria zizanoides* | 2.25 | |
| Sole Coconut | 1.88 | |
| Coconut + *Alpina galangal* | 3.20 | Mohandas, |
| Coconut + *Aloe vera* | 3.80 | S. 2011 |
| Coconut + *Ocimum sanctum* | 3.00 | |
| Coconut + *Cymbopogan flexuosus* | 2.10 | |

| Agro-forestry system | B/C ratio | References |
|---|---|---|
| Coconut + *Pogestemon patchouli* | 2.10 | |
| Peach + Grewia + Setaria + Withania | 3.07 | Verma and Thakur 2011 |
| Peach + Morus + Setaria + Withania | 3.12 | |
| Peach + Setaria + Withania | 3.23 | |
| Grewia + Setaria + Withania | 3.57 | |
| Morus + Setaria + Withania | 3.87 | |
| Sole Withania | 3.06 | |
| Coconut + Guava + *Aloe vera* | 3.54 | Bari, M. S. and Rahim, M. A. 2012 |
| Coconut + Lemon + *Aloe vera* | 2.82 | |
| Open *Aloe vera* | 1.65 | |
| Coconut + Guava + Asparagus | 3.21 | |
| Coconut + Lemon + Asparagus | 3.08 | |
| Open Asparagus | 1.87 | |
| Coconut + Guava + Misridana | 5.06 | |
| Coconut + Lemon + Misridana | 3.83 | |
| Open Misridana | 2.33 | |
| Amla (*Emblica officinalis*) + Aswagandha | 1.60 | Madhavi *et al.*, 2012 |
| Terminalia (*Terminalia chebula*) + Aswagandha | 9.45 | |
| Sole Aswagandha | 1.07 | |
| *Emblica officinalis* + *Andrographis paniculata* | 2.47 | Madhavi *et al.*, 2012 |
| *Terminalia chebula* + *Andrographis paniculata* | 11.11 | |
| Sole *Andrographis paniculata* | 1.86 | |
| Peach + Grewia + Setaria + Ocimum | 3.07 | Thakur and Verma 2012 |
| Peach + Morus + Setaria + Ocimum | 3.12 | |
| Peach + Setaria + Ocimum | 3.23 | |
| Grewia + Setaria + Ocimum | 3.58 | |
| Morus + Setaria + Ocimum | 3.88 | |
| Sole Ocimum | 3.08 | |
| Coconut + Guava + *Aloe vera* | 3.54 | Bari and Rahim 2012a |
| Coconut + Lemon + *Aloe vera* | 2.82 | |
| Sole *Aloe vera* | 1.65 | |
| Coconut + Guava + *Asparagus racemosus* | 3.21 | Bari and Rahim 2012b |
| Coconut + Lemon + *Asparagus racemosus* | 3.08 | |
| Sole *Asparagus racemosus* | 1.87 | |
| Coconut + Guava + *Kaempferia angustifolia* | 5.06 | |
| Coconut + Lemon + *Kaempferia angustifolia* | 3.83 | |
| Sole *Kaempferia angustifolia* | 2.33 | |
| Gamhar + Pineapple | 1.65 | Nayak *et al.*, 2014 |
| Gamhar + *Aloe vera* | 1.30 | |
| Gamhar + Kalmegh | 1.56 | |
| Gamhar + Mango ginger | 1.37 | |

| Agro-forestry system | B/C ratio | References |
|---|---|---|
| Sole Pineapple | 1.58 | |
| Sole *Aloe vera* | 1.16 | |
| Sole Kalmegh | 2.27 | |
| Sole Mango ginger | 2.19 | |
| Manilkara achras + *Jatropha curcas* + *Ocimum sanctum* | 1.71 | Solanki *et al.*, 2014 |
| Manilkara achras + *Jatropha curcas* + *Andrographis paniculata* | 1.46 | |
| Manilkara achras + *Jatropha curcas* + *Mentha arvensis* | 1.42 | |
| Sole *Ocimum sanctum* | 1.69 | |
| Sole *Andrographis paniculatasole* | 1.28 | |
| Sole *Mentha arvensis* | 1.26 | |
| Teak + *Ocimum tenuiflorum* | 1.01 | Kumar *et al.*, 2015 |
| Teak + *Ocimum gratissimum* | 1.24 | |
| Teak + *Ocimum basilicum* | 1.13 | |
| Sole Teak + Patchouli | 1.94 | Kumar *et al.*, 2016 |
| Sole Patchouli100% FYM (20 t/ha) | 1.87 | |
| Teak + Patchouli100% FYM (20 t/ha) | 2.14 | |
| Sole Patchouli100% VC (10 t/ha) | 2.07 | |
| Teak + Patchouli100% VC (10 t/ha) | 2.32 | |
| Sole Patchouli100% N C (0.5–1.0 t/ha) | 1.87 | |
| Teak + Patchouli100% N C (0.5–1.0 t/ha) | 2.04 | |
| Sole Patchouli 50% FYM + 50% NC | 1.79 | |
| Teak + Patchouli 50% FYM + 50% NC | 2.10 | |
| Sole Patchouli 50% VC + 50% NC | 1.91 | |
| Teak + Patchouli 50% VC + 50% NC | 2.21 | |
| Sole Patchouli 50% FYM + 50% VC | 2.01 | |
| Teak + Patchouli 50% FYM + 50% VC | 2.35 | |
| Drumstick + Safedmusli, Kalam khet (Gujarat) | 1.42 | Mehta *et al.*, 2017 |
| Drumstick + Safedmusli, Melghat (Maharashtra) | 2.41 | |
| Drumstick + Safedmusli, JNKVV (M.P.) | 1.61 | |
| Drumstick + Safedmusli, Satpuda (Maharastra) | 1.53 | |
| Drumstick + Safedmusli, Sahyadri (Maharashtra) | 2.76 | |
| Drumstick + Safedmusli, Bhuvadi (Gujarat) | 2.44 | |
| Drumstick + Safedmusli, Rambhas chikar (Gujarat) | 2.47 | |
| Drumstick + Safedmusli, Pratapgarh (Rajasthan) | 2.23 | |
| Drumstick + Safedmusli, SFRI (M.P.)) | 1.64 | |
| Drumstick + Safedmusli, Dhar (M.P.) | 1.60 | |
| Sole *Cymbopogon* spp. | 2.97 | Mohanty *et al.*, 2017 |
| *M. dubia* (2×2 m) + *Cymbopogon* spp. | 3.86 | |
| *M. dubia* (3×3 m) + *Cymbopogon* spp. | 4.41 | |
| *M. dubia* (3×2 m) + *Cymbopogon* spp. | 4.33 | |

| Agro-forestry system | B/C ratio | References |
|---|---|---|
| *M. dubia* (4×4 m) + *Cymbopogon* spp. | 3.14 | |
| *M. dubia* (4×2 m) + *Cymbopogon* spp. | 3.95 | |
| Sole *M. dubia* (2×2 m) | 0.25 | |
| Sole *M. dubia* (3×3 m) | 5.44 | |
| Sole *M. dubia* (3×2 m) | 1.59 | |
| Sole *M. dubia* (4×4 m) | 4.02 | |
| Sole *M. dubia* (4×2 m) | 1.64 | |
| Sole fennel | 1.24 | Rathee *et al.*, 2017 |
| Poplar + fennel (5×4 m) | 1.27 | |
| Poplar + fennel (10×2 m) | 1.37 | |
| Poplar + fennel (18×2×2 m) | 1.66 | |
| Sole ajwan | 0.70 | |
| Poplar + ajwan (5×4 m) | 1.10 | |
| Poplar + ajwan (10×2 m) | 1.20 | |
| Poplar + ajwan (18×2×2 m) | 1.32 | |
| Mango + Ginger | 1.87 | Gupta, M. 2018 |
| Mango + Turmeric | 2.29 | |
| Mango + Stevia | 2.30 | |
| Mango + Kalmegh | 3.00 | |
| *M. dubia* + *A. vera* (2×2 m) | 3.82 | Jilariya *et al.*, 2019 |
| *M. dubia* + *A. vera* (3×2 m) | 3.26 | |
| *M. dubia* + *A. vera* (4×2 m) | 3.25 | |
| Sole *A. vera* | 1.32 | |
| Sole *M. dubia* (2×2 m) | 2.66 | |
| Sole *M. dubia* (3×2 m) | 3.59 | |
| Sole *M. dubia* (4×2 m) | 3.71 | |

## Conclusion

Medicinal & Aromatic plants can be intercropped with tree species depending on the size, shape and intensity of its canopy shade, tree spacing and management, especially thinning, pruning of branches and nature of the MAPs. Shade tolerant spp. and rhizomatic MAPs can be grown on a longer term basis in broadly spaced plantations. Most of medicinal and aromatic plants are found to be shade tolerant species; therefore agroforestry system offers a convenient strategy for promoting their cultivation and conservation. Medicinal and Aromatic plants based agroforestry system through, reduce the pressure on the dwindling resources, to obtain sustainable and regular supply of wood, fibre, fruit and medicinal and aromatic products to catering the industrial demand, to improve the microclimate by lowering the surface soil temperature, to improve the soil physical, chemical and biological properties, to reduce the soil erosion and degradation, and finally to increase the farm income or doubling farmers income.

## References

Bari MS and Rahim MA 2012 Economic Evaluation and Yield Performance of Some Medicinal Plants in Coconut Based Multistoried Agroforestry Systems. *The Agriculturists,* **10**(1) 71–80.

Bari MS and Rahim MA 2012a Bio-economic evaluation of Aloe vera in coconut based multistoried agroforestry systems. *J. Agrofor. Environ.* 3(2) 61–63.

Bari MS and Rahim MA 2012b Economic evaluation and yield performance of some medicinal plants in coconut based multistoried agroforestry systems. *The Agriculturists* **10**(1) 71–80.

Basavaraju TB and Nanjappa HV, Umesha K, Vasundhara M and Arulraj S 2011 Intercropping of medicinal and aromatic plants in coconut gardens. *Journal of Plantation Crops* **39**(2) 299–304.

Gupta M 2018. Component interactions in mango based agroforestry system in the sub-tropics of Jammu and Kashmir. Ph.D. thesis, Division of Agroforestry, Faculty of Agriculture, Sher-e-Kashmir University of Agricultural Sciences and Technology of Jammu, Jammu–180009 (J & K).

Jilariya DJ, Thakur NS, Singh N and Gunaga RP 2019 Economics of cultivation of *Melia dubia* Cav.–*Aloe vera* L. silvi–medicinal model. *Indian J. of Agroforestry* **21**(2) 35–40.

Kumar D, Bijalwan A, Kalra A and Dobriyal MJR 2016 Effect of shade and organic manure on growth and yield of patchouli [*Pogostemon cablin* (blanco) benth.] under teak (*Tectona grandis* L. F.) based agroforestry system. *Indian Forester* **142**(11) 1121–1129.

Kumar M, Thakur NS and Hegde HT 2015 Growth, herb yield and financial flows from Ocimum *spp.* Intercroppedunder teak (*Tectona grandis*L.f.)–*Ocimum* spp. based silvi–medicinal system in Gujarat, India. *International Journal of Innovative Horticulture* **4**(2) 113–118.

Kumar RD, Sreenivasulu GB, Prashanth SJ, Jayaprakashnarayan RP, Nataraj SK and Hegde NK 2010 Performance of safed musli (*Chlorophytum borivilianum.* L) in tamarind plantation as intercrop and as sole crop in open area. *International Journal of Agricultural Science* **6**(1) 359–360.

Madhavi LA, Srinivasa RM, Joseph B, Siva SA and Chandrasekhar RP 2012 Performance of medicinal herbs in tree based cropping system under dryland conditions. *Journal of Pharmacognosy* **3**(2) 92–95.

Mehta AA, Tandel MB, Patel DP, Behera LK, Prajapati DH and Jadeja DB 2017 Yield Performance of *Chlorophytum borivilianum* Sant. & *Fernand accessions* in Moringa based Agroforestry system. *International Journal of Agriculture Sciences* **9**(10) 3976–3979.

Mohandas S 2011 Prospects of intercropping medicinal and aromatic plants in coconut garden. *Madras Agriculture Journal* **98**(1 3) 82–83.

Mohanty S, Thakur NS, Gunaga RP, Dobriyal MJ and Desai BS 2017 Value Addition in *Cymbopogon spp.* to Enhance the Financial Flows from *Cymbopogon* spp.–*Melia dubia* Cav. based Silvi–medicinal and Sole Cropping Systems. *Indian Journal of Ecology* **44**(6) 812–816.

Nayak MR, Behera LK, Mishra PJ and Bhola N 2014 Economics and yield performance of some short duration fruit and medicinal crops under agri-silvicultural system in rainfed uplands of Odisha. *J. Appl.* and *Nat. Sci.* **6**(1) 274–278.

Rathee P, Kaushik N, Khajuria S, Singh P and Manjeet 2017 Performance of Fennel and Ajwan as Intercrops under Different Spacings of Poplar Plantations in North–Western, India. *Int. J. Pure App. Biosci.* **5**(1) 901–907.

Solanki VK, Jadeja DB and Tandel MB 2014 Performance of herbal medicinal crops under sapota-jatropha based three-tier agroforestry system. *Internat. J. agric. Sci.* **10**(1) 267–271.

Thakur NS and Verma KS 2012 Financial flows from sacred basil (*Ocimum sanctum*) based agroforestry land use systems in mid hills of Western Himalayas. *Indian Forester* **138**(7) 638–645.

Thakur NS, Attar SK and Chauhan RS 2017 Horti–Medicinal Agroforestry Systems: A Potential Land Use for Commercial Cultivation of Medicinal and Aromatic Plants. In Agroforestry for Increased Production and Livelihood Security, 163–183 pages.

The Hindu. Medicinal plants in good demand, 8 December 2019, Kerela. https://www.thehindu.com/news/national/kerala/medicinal-plants-in-gooddemand/article 30239 519.ece

Tripathi P 2012 Effect of organic manures on yield and biomass production of medicinal and aromatic plants under peach based agroforestry system. Ph.D. thesis, Dr. Y. S. Parmar University of Horticulture and Forestry, Nauni, Solan (H. P.), 163 page.

Verma KS and Thakur NS 2010 Economic analysis of ashwagandha (*Withania somnifera* L.) based agroforestry land use systems in mid hills Western Himalayas. *Indian Journal of Agroforestry* **12**(1) 62–70.

Verma KS and Thakur NS 2011 Economic appraisal of kewach (*Mucuna pruriens*) based agroforestry land–use systems in mid hills of Western Himalayas. *Journal of Non-Timber Forest Products* **18**(1) 47–54.

CHAPTER - 17

# Agroforestry based Integrated Farming System

## INTEGRATED FARMING SYSTEM (IFS)

Integrated farming system is a common term used to explain a more integrated approach towards farming compared to conventional monoculture approach. According to Lal and Miller (1990), farming system is a resource management strategy to achieve economic and sustained agricultural production to meet diverse requirements of farm livelihood while preserving resource base and maintaining a high level of environment quality. It refers to a system that integrates livestock (dairy, poultry, fish etc.) with crop production and sometimes can also be called as integrated bio-systems. The enterprises are selected and managed in such a way that the "waste" from one component becomes an input for another part of the system. This reduces the cost of production and thus income is improved. By utilizing waste as a resource, which on accumulation may lead to pollution, it also helps eliminating waste from the system.

In the last few decades, "modern" technologies have been widely used to enhance the productivity per unit of land to meet the increasing food demand for the ever-increasing global population. Due to non-judicial over use of toxic agrochemicals, our food and ecosystems have been poisoned. With exploding population and rapid depletion and degradation of natural resource base, sustainable agriculture has assumed very great significance. The task of meeting the needs of the present generation, without eroding the ecological assets of future generations is receiving top priority by the planners. In recent years, food security, employment and income generation, as well as resource conservation and environment protection have emerged as major world concerns. Especially developing countries struggling with these issues have to deal with the dual burden of climate change and globalization. Thus, developing countries around the world are promoting sustainable development through sustainable agricultural practices which will help them in addressing socio-economic as well as environmental issues simultaneously.

The concept of sustainability is the most important element in development of integrated farming systems. Systems that improves biological interactions between components, focuses on diversity of crops as well as enterprises rather than obtaining simple crop yield (Olsen *et al.* 2000), and consider tree management as part of the farm strategy (Vandemeer *et al.* 1998, Rietveld and Francis 2000) are recognized as sustainable alternatives to conventional agriculture. Annual crops are most affected by quantity and distribution of rainfall and the frequency of drought is higher in arid region. In India, over 68% of area is vulnerable to drought, out of which 33% area receiving <750 mm rainfall is chronically drought prone. To minimize the risk and uncertainty associated with agriculture, farmers all over the world practice farming system in one form or another. Adopting a suitable mix of enterprises is the first step towards preparedness to risks as no single enterprise can impart stability in income, especially in fragile environments. Traditional farming systems represent a crop– livestock– tree/shrub– human continuum (Tanwar and Tewari, 2017) or crop, livestock, aquaculture, agroforestry, forestry, agri-horticulture and sericulture (Singh, 2004).

Through adoption of integrated farming system and applying modern farming practices, the problems of food security and global warming mitigation should definitely be solved.

## Definition of integrated farming

Different scientists have defined farming system. However, many definitions, in general, convey the same meaning that it is strategy to achieve profitable and sustained agricultural production to meet the diversified needs of farming community through efficient use of farm resources without degrading the natural resource base and environmental quality. Relatively recent definitions include:

- Farming system is a decision-making unit comprising the farm household, cropping and livestock system that transform land, capital and labour into useful products that can be consumed or sold (Fresco and Westphal, 1988).
- Farming system is a set of agro economic activities that are interrelated and interact with themselves in a particular agrarian setting. It is a mix of farm enterprises to which farm families allocate its resources in order to efficiently utilize the existing enterprises for increasing the productivity and profitability of the farm. These farm enterprises are crop, livestock, aquaculture, agro forestry and agri-horticulture (Sharma *et al.*, 1991).
- Farming system is a mix of farm enterprises such as crop, livestock, aquaculture, agro forestry and fruit crops to which farm family allocates its resources in order to efficiently manage the existing environment for the attainment of the family goal. (Pandey *et al.*, 1992).
- Farming system represents an appropriate combination of farm enterprises (cropping systems horticulture, livestock, fishery, forestry, poultry) and the means available to the farmer to raise them for profitability. It interacts adequately with environment without dislocating the ecological and socioeconomic balance on one hand and attempts to meet the national goals on the other (Jayanthi *et al.*, 2002).

### Advantages of farming system

1. Recycling and utilization of resources available in the farm
2. Higher and sustainable production
3. Maximum possible return resulting in higher profitability
4. Adequate and year-round employment opportunities
5. Food and nutritional security
6. Environmental Safety
7. Adoption of New Technology
8. Energy saving
9. Solving Fuel and Timber Crisis
10. Agro-industries
11. Increasing Input Efficiency

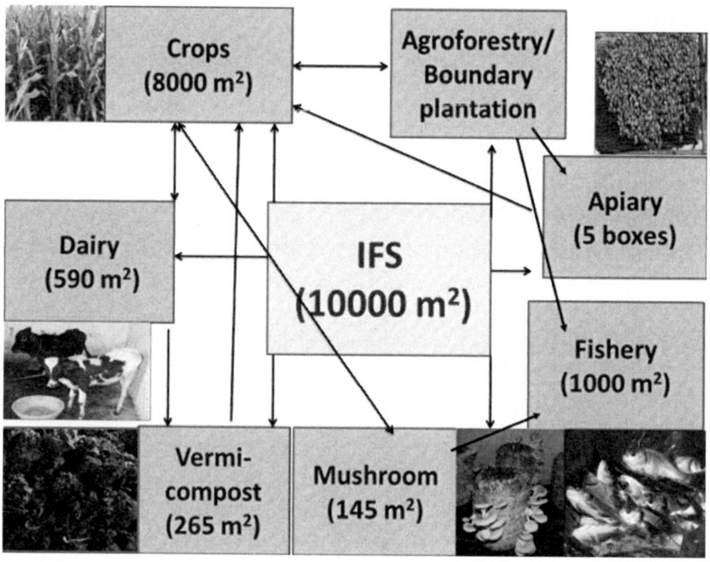


**Fig. 17.1:** 1 ha agroforestry based IFS model in BAU, Ranchi, Jharkhand

Different Promising Integrated farming system which suitable on the basis of income/year/ha for different regions of India.

**Table 17.1:** Promising IFS for Indo-Gangetic plains

| Integrated Farming System | Income (Rs./ha/yr) | Reference |
|---|---|---|
| Rice–wheat | 81,200 (Gross) | Gill, 2004 |
| Rice–wheat + dairy | 1,54,000 (Gross) | (Punjab) |
| Rice–wheat + fish + piggery | 1,13,200 (Gross) | |
| Crop alone | 4,615 | Singh *et al.*, 1993 |
| Crop + crossbred cow | 20,581 | (Haryana) |
| Crop + buffalo | 6,218 | |

| Integrated Farming System | Income (Rs./ha/yr) | Reference |
|---|---|---|
| Sugarcane–wheat | 41,017 | Singh, 2004 |
| Sugarcane–wheat + dairy | 47,737 | (Uttar Pradesh) |
| Crops alone | 66,371 | Singh *et al.*, 2006 |
| Crop + dairy | 1,03,615 | (Uttar Pradesh) |
| Crop + dairy + horticulture | 1,07,467 | |
| Crop + dairy + apiary | 1,34,382 | |
| Crop + dairy + vermicomposting | 1,39,472 | |
| Crop + fish + duck + goat | 159485 | Kumar *et al.*, 2012 (Bihar) |

**Table 17.2:** Promising IFS for central India

| Integrated Farming System | Income (Rs./ha/yr) | Reference |
|---|---|---|
| Crops alone | 24,093 | Tiwari *et al.*, 1999 |
| Mixed farming + 2 cow | 37,668 | (Madhya Pradesh) |
| Dairy ( 2 cows) + 15 goats + 10 poultry + 10 duck + fish | 44,913 | |
| Cotton + groundnut | –92 | Shelke *et al.*, 2001 |
| Blackgram–onion–maize + cowpea | 1,304 | (Maharashtra) |
| Crop + dairy + sericulture | 3,524 | |
| Crop + dairy | 5,121 | |

**Table 17.3:** Promising IFS for southern India

| Integrated Farming System | Income (Rs./ha/yr) | Reference |
|---|---|---|
| Rice-rice | 15,299 | Shanmugasundaram |
| Rice-rice-Azolla/Calotropis + fish | 17,488 | and Baluswamy, 1993 (Tamilnadu) |
| Rice-rice-blackgram | 8,312 | Shanmugasundaram |
| Rice-rice-cotton + maize | 15,009 | *et al.*, 1995 |
| Rice-rice-cotton + maize + poultry/fish | 17,209 | (Tamilnadu) |
| Rice | 22,971 | Baluswamy *et al.*, |
| Rice + fish | 28,569 | 2003 (Tamilnadu) |
| Rice + Azolla + fish | 31,788 | |
| Cropping alone | 36,190 | Jayanthi *et al.*, 2001 |
| Cropping + fish + poultry | 97,731 | (Tamilnadu) |
| Cropping + fish + pigeon | 98,778 | |
| Cropping + fish + goat | 1,31,118 | |

| Integrated Farming System | Income (Rs./ha/yr) | Reference |
|---|---|---|
| Rice-rice-rice-fallow-pulses | 13,790 | Ganeshan *et al.*, 1990 |
| Rice-rice-rice-fallow-cotton + maize + duck cum fish | 24,117 | (Tamilnadu) |

**Table 17.4:** Promising IFS for coastal region

| Integrated Farming System | Income (Rs./ha/yr) | Reference |
|---|---|---|
| Cashew | 36,330 | Manjunath and Itnal, |
| Coconut + forage + dairy | 32,335 | 2003 (Goa) |
| Rice-brinjal (0.5ha) + rice cowpea (0.5ha) + mushroom + poultry | 75,360 | |
| Coconut alone | 10,400 | Parthasaraty, *et al.*, |
| Coconut + ginger | 36,590 | 2004 (Kerala) |
| Coconut + clove | 38,500 | |
| Rice alone | 19,210 | Korikanthimath and |
| Rice-groundnut + mushroom + poultry | 60,650 | Manjunath, 2009 |
| Rice-cowpea + mushroom + poultry | 73,430 | (Goa) |
| Rice-brinjal + mushroom + poultry | 77,310 | |
| Rice-sunhemp + mushroom + poultry | 52,750 | |
| Agriculture + poultry | 18,000 | Viroji Rao and Bose, |
| Agriculture + sheep | 43,000 | 2002 (Andhra Pradesh) |
| Agriculture + sericulture | 10,000 | |

## Agroforestry system in Integrated farming system

The integration of forestry and other agricultural activities on a farm, or simply raising trees on farm, also known as farm forestry, is an age-old practice in many parts of the world. It is a tested and well-established technology to meet the economic and environmental goals. Since the interactive benefits of agro-forestry system are not immediately visible, a holistic interpretive framework for on-farm activities such as Farming System Research methodology is necessary to understand or suggest adaptations in this system (Jones and Price, 1985).

Integrated farming system follow agro-forestry settings with crop production, livestock production like dairy/goat/sheep rearing, backyard poultry, vermiculture, and fodder production in intercropping with fruit trees, bio bunds and live fencing, make considerable amount of profit. Agroforestry, tree intercropping, live furrow, live fencing in between orchard, green fodder production, jasminum pasture might be adopted in integrated farming system.

**Tree intercropping method**: Fodder tree species are planted in specified interval and in between the spaces Agriculture crops can be grown this method is called as tree intercropping system. On this land tree fodders can be planted in 10 m gap *viz.*, Subabul and glyciridia can be planted. In this method we can get 10 to 11 tons of green fodders which will be sufficient to feed 15 to 18 sheeps.

**Setting bio-bunds:** In general, to prevent soil erosion and conserving moisture, the bunds of 1 to 2 feet height are made and grasses like kolukkattaipul, guinea grass are planted on the bunds. On both sides in 10 m gap, Subabul or Gliricidia seedlings can be planted. These plants after developing well should be cut at a height of 6 feet from the ground. The new branches will grow much faster and get high tree leaves. From approximately 100 m length 500 to 1000 kg greens can be produced.

**Bio-fencing:** Instead of setting up of iron fencing, planting fodder trees on the boundaries as bio fence. Surrounding the land with larger growing trees of wood species like udhiyan, thespesia, kodukkapuli could be planted at 4 to 5 m intervals. In the gap between these trees small trees like Subabul, Gliricidia can be grown. These trees should be pruned at 1.5 m height to generate new lateral branches that can be used as feed for animals. From the bio fence we can get 3 or 4 tons of fodder annually for livestock.

**Fodder production in coconut garden:** Coconut trees are usually planted at a spacing of 8 m. In general, in between coconut trees the land is not used properly and left as fallow land. Because of this soil fertility is reduced in coconut plantation. Whatever the lateral extent of the coconut tree leaves goes similarly the length of the root also extended. Therefore, the gap between the two trees, small trees such as subabul, legume fodder varieties such as kalappakkoniyam and stylo can be cultivated to increase the fodder production. Planting stylo or any other leguminous fodder in coconut plantation also helps to fix the nitrogen and increase the coconut productivity and also supply protein rich fodder to the livestock. In this method, from an acre of coconut garden we can get about 15 tons fodder.

**Fodder production between orchard trees:** Fruit trees are often planted in the spacing of 10 m × 10 m or 12 m × 12 m in the orchards. Generally, roots of fruit trees spread horizontally, laterally and deeply based on water requirement. In between fruit trees soil surfaces can be utilized for raising cattle fodders. During heavy rainfall season, the rain causes soil erosion in the fallow interspacing area of fruit trees. Forage crops or any plant sown in the fallow interspace will serve as a cover crop and protect soil erosion.

**Mullai pasture land development:** In irrigated lands, when cumbu-napier grass is grown, shrubs like subabul, Gliricidia, Mulberry, Agathi planted in 2 × 2 $m^2$ spacing as intercrop and its leaves harvested and given as fodder.

**Sorghum + Stylo:** On some lands cultivation is done only once a year due to unavailability of moisture. For example after sorghum and pearl millet cultivation the particular land kept as fallow and due to this soil erosion occurs. Instead of this practice, after sowing of main crops, stylo seeds can be sown @ 14 kg/ha. Sorghum and cumbu are fast growing in nature whereas stylo is slow in growth. After harvest of grain crops like sorghum and cumbu, the stylo receives direct sunlight and show profound growth. If this crop harvested after attaining full maturity the seeds will fall on the land and it will germinate in the next year monsoon season. Hence once sown, stylo gives forage for next 2–3 years even without grain crops cultivation. Stylo also fixes the atmospheric nitrogen in the soil which helps increase the yield of sorghum.

Therefore, those who are all involved in agriculture and livestock farming practices facing input availability and labor scarcity problems can get increased productivity and additional income through agroforestry as integrated farming system is a better approach to get sustained income.

## Components of Agroforestry based Integrated Farming Systems (Malik *et al.*, 2018)

A. **Livelihood security model:** (Fruit/ vegetable /fodder /fiber /cereals, oil seeds and pulses etc. based)

*Example:*

(i) Horticultural trees (Guava /Mango /Litchi /Aonla etc.) + Agricultural crops (Pea /Beans /Tomato / Cauliflower /Chilli, etc.)

(ii) Nitrogen fixing tree (Subabool /Sisham) + Agricultural crops (Upland paddy/Groundnut/ Arhar etc.)

(iii) Fuel wood tree (Chakundi/Subabul) + Agricultural crops (Upland paddy /Groundnut /Arhar etc.)

(iv) Trees on field boundary (Gamhar/Sisham/Karanj) + Agricultural crops/ fodder crops (Berseem/Guinea grass/Oat/Dinanath grass/Lucerne etc.)

B. **Nutritional & Health Security model:** (Fruit/vegetable/milk/Meat/egg etc. based)

*Examples:*

(i) In pond -Pisiculture, duckery, etc.

On bank -Horticultural trees (Guava/Mango/Banana/Citrus/Moringa /Papaya etc.) + Livestock (Goat/Poultry/Pig/Duck etc.)

(ii) Around pond-Agricultural crops (Grain crops/Fodder crops/vegetable crops) + Forest crops (Gamhar/Subabul/Moringa etc.)/Fruit trees (Guava /Mango/banana/Citrus/ papaya etc.)

C. **Income generation model:** (Cash crops/Medicinal crops/Aromatic crops/Horticultural crops/livestock)

*Examples:*

(i) Cash crops (Capsicum/Brinjal/Tomato/Peas/Groundnut etc.) + Fruit trees (Madhuka /Jackfruit/Ber/Anola/lime etc.)/ Fuel trees (Chakundi/ Subabul etc.)/ Timber trees (Gamhar/Sisham etc.)

(ii) Agricultural crops (Maize/Groundnut/Arhar etc.) horticultural crops (Marigold /Gladulai/Roses etc.)

(iii) Medicinal/Aromatic (Lemon grass/Tulsi/Stevia/Safedmusli etc.) + Timber tree (Gamhar/teak/Sisham etc.) Floriculture (Marigold/Gladulai/Roses etc.)

(iv) Pond model (Milk, egg, and meat production model) Rearing of Goat/poultry/Pig/Duck etc.

**Fig. 17.2:** Agroforestry based IFS (BAU, Ranchi, Jharkhand)

## Economics of Agroforestry System in Integrated farming System

The integration of 7 different enterprises namely, crop+fish+goat+Vermicompost+ fruit production + spice production + agro forestry obtained the net return to the tune of Rs. 2, 30,329 annually with the Benefit Cost Ratio (BCR) of 1.07:1 and also reported the maximum per cent contribution of the enterprise is the fish production (68.53 per cent) followed by vermicomposting (9.90 per cent), spices (8.46 per cent) and animal production (7.40 per cent). The BCR was found to be highest for the spice production (1.83:1) after fishery (2.25:1) followed by the vermincomposting (1.45:1) (Tripathi *et al.*, 2010).

**Table 17.5:** Economic viability of Integrated Farming System Research models developed in different states of the country

| State | Prevailing system | Net return | Integrated Farming System | Net returns |
|---|---|---|---|---|
| Tamilnadu | Rice-rice-blackgram | 8,312 | Rice-rice-cotton + maize | 15,009 |
| | | | Rice-rice-cotton + maize + poultry/fish | 17,209 |
| | Rice-rice | 15,299 | Rice-rice-Azolla/Calotropis + Fish | 17,488 |
| | Rice-rice-rice-fallow-Pulses | 13,790 | Rice-rice-rice-fallow-cotton + maize + duck cum fish | 24,117 |
| | | | Cropping + fish + poultry | 97,731 |
| | Cropping alone | 36,190 | Cropping + fish + pigeon | 98,778 |
| | | | Cropping + fish + goat | 13,1118 |
| | Rice | 22,971 | Rice + fish<br>Rice + Azolla + fish | 28,569<br>31,788 |
| Goa | Cashew | 36,330 | Coconut + forage + dairy<br>Rice-brinjal (0.5 ha) + Rice-cowpea (0.5 ha) + mushroom + poultry | 32,335<br>75,360 |
| Madhya Pradesh | Arable farming | 24,093 | Mixed farming + 2 cow<br>Dairy (2cows) + 15 goats + 10 poultry + 10 duck + fish | 37,668<br>44,913 |
| Maharashtra | Cotton (K) + Groundnut (S) | (-) 92 | Blackgram (K) - Onion (R)-<br>Maize + cowpea Crop + dairy<br>+ sericulture Crop + dairy | 1,304<br>3,524<br>5,121 |
| Uttar Pradesh | Crops (Sugarcane-wheat) | 41,017 | Crops (Sugarcane + wheat) + dairy | 47,737 |

| State | Prevailing system | Net return | Integrated Farming System | Net returns |
|---|---|---|---|---|
| | | | Rice-fish (pit at the center of the field) - poultry (reared separately) | 62, 977 |
| Karnataka | Rice-rice system | 21599 | Rice-fish (pit at one side of the field) - poultry (shed on fish pit) | 49, 303 |

**Case study 1:** The IFS model at Phadiarkhar Farm was initiated during rabi 2010–11.

**Table 17.6:** Summary of the component wise costs, gross returns and net returns in on station IFS model (2012–13)

| Component wise summary | | | | | |
|---|---|---|---|---|---|
| **Component** | **Year of start** | **Net Area** | **Total Cost (Rs./ha)** | **Gross Revenue (Rs./ha)** | **Net returns (Rs./ha)** |
| Cropping systems | 2010 | 0.65 | 24965 | 49425 | 24460 |
| Horticulture | 2010 | 0.175 | 21890 | 26841 | 4951 |
| Fruit crops | 2010 | — | — | — | — |
| Floriculture | — | — | — | — | — |
| Agro-forestry (Leucenia Plants ) | 2012 | Border plantation | — | — | — |
| Livestock | — | — | — | — | — |
| Dairy animals | 2010 | | 83373 | 139473 | 56100 |
| Goats/ Sheep | — | — | — | — | — |
| Poultry/ Ducks | — | — | | — | — |
| Others (Fodder Block) | 2010 | 0.10 | 4813.1 | 3890 | -923 |
| Fisheries | — | — | — | — | — |
| Apiary | — | — | — | — | — |
| Others/ (specify) | — | — | — | — | — |
| All components | — | 0.925 | 135041 | 219629 | 84588 |

*Source:* Lal *et al.*, 2018

**Case study 2:** Development of Region Specific Horticulture based Integrated Farming System Models of Sirohi District of Rajasthan (Bhardwaj and Vyas, 2015)

IFS model consisting of three layers (lime + papaya + marigold or pomegranate + papaya + marigold) orchard, vegetables and off-season vegetables for regular and sustainable income generation from the farm. The data shows that those farmers adopting model 7 (Crop + cash crop + vegetable + flower + fruit production with retail marketing of fruit and vegetables) earned Rs. 5.30 lakh year$^{-1}$ with maximum B:C ratio (4.38) and employment opportunity of 1800 man days year$^{-1}$ was generated at farmers field from 2 ha cultivable land, whereas under model–1 (crop production) the farmers earned annually Rs. 1.22 lakh with lowest benefit cost ratio (2.03).

**Table 17.7:** Horticulture based farming system model

| S.No | Models | Types of Models |
|---|---|---|
| 1. | Models–1 | Crop production |
| 2. | Models–2 | Cash crop production |
| 3. | Models–3 | Vegetable production |
| 4. | Models–4 | Crop + cash crop + vegetable production |
| 5. | Models–5 | Crop + cash crop + vegetable + fruit production |
| 6. | Models–6 | Crop + cash crop + vegetable + flower + fruit production |
| 7. | Models–7 | Crop + cash crop + vegetable + flower + fruit production with retail marketing |

**Table 17.8:** Region specific integrated farming system models for partially irrigated arid condition in 2.0 hectare

| Models | Major component | Cost of cultivation (in lakh) | Gross return (in lakh) | Net return (in lakh) | Benefit –cost ratio |
|---|---|---|---|---|---|
| Models- 1 | ***Rabi***–wheat, mustard, barley, gram etc. ***Kharif***–Maize, bajra, green gram, cluster bean etc. | 0.60 | 1.82 | 1.22 | 2.03 |
| Models- 2 | Fennel- 0.5ha Castor –0.5ha Cotton- 0.5 ha Other crop- 0.5 ha | 0.78 | 2.84 | 2.06 | 2.64 |
| Models- 3 | ***Rabi***–Tomato, brinjal, colecrop, pea ***Kharif***- Okra, cucurbits | 1.68 | 6.80 | 4.12 | 3.04 |

| Models | Major component | Cost of cultivation (in lakh) | Gross return (in lakh) | Net return (in lakh) | Benefit –cost ratio |
|---|---|---|---|---|---|
| | ***Zaid***– Okra, chilli, cucurbits | | | | |
| Models– 4 | Crop–0.6 ha<br>Cash crop–0.7ha<br>Vegetable–0.7ha | 1.02 | 2.85 | 2.91 | 2.86 |
| Models– 5 | Crop–0.5 ha<br>Cash crop–0.5 ha<br>Vegetable –0.5ha<br>Fruit plants–0.5ha | 0.90 | 4.32 | 3.33 | 3.75 |
| Models– 6 | Crop–0.4 ha<br>Cash crop–0.4 ha<br>Vegetable –0.4 ha<br>Fruit plants–0.4ha<br>Flower –0.4 ha | 1.05 | 5.07 | 4.01 | 3.80 |
| Models– 7 | Crop–0.4 ha<br>Cash crop–0.4 ha<br>Vegetable –0.4 ha<br>Fruit production–0.4ha<br>Flower –0.4 ha with retail marketing | 1.20 | 6.50 | 5.29 | 4.38 |

Growing of trees with agricultural crops is an age old practice in arid semi arid and other regions. Trees not only give shade but in virtue of their perennial nature provide a steady supply of food and/or income throughout the year, check degradation and maintain soil fertility, diversify income sources and stabilize it, enhances efficiency of soil nutrients, water and radiation and provide regular employment.

## Conclusion

These examples demonstarate the potential of agroforestry in improving economic returns and reducing vulnerability in agriculture system. Thus, agroforestry is the most important component in an integrated farming system to achieve sustainability.

## References

Agro Forestry and Integrated Farming System, TNAU, Agritech Potal, 2016.

Bhardwaj RL and Vyas L 2015 Development of Region Specific Horticulture based Integrated Farming System Models with Crop Diversification for Sustainable Livelihoods and Nutritional Security of Tribal of Sirohi District of Rajasthan: A Case Study. *Advances in Social Research* 1(1) 17–29.

Fresco LO and Westphal E 1988 A hierarchical classification of farm systems. *Experimental Agriculture* **24** 399–419.

Jayanthi C 2002 Sustainable Farming System and lowland farming of Tamil Nadu. IFS Adhoc. scheme. Completion Report.

Jones JR and Price N 1985 Agroforestry: An Application of the Farming Systems Approach to Forestry. *Society for Applied Anthropology* **44**(4) 322–331. https://www.jstor.org /stable/44126265.

Lal M, Patidar J, Kumar S. and Patidar A 2018 Different integrated farming system model for irrigated condition of India on basis of economic assessment: A case study: A review. *International Journal of Chemical Studies* **6**(4) 166–175.

Lal R and Miller FP 1990 Sustainable farming for tropics. In: Singh, R.P. (Ed.) Sustainable agriculture: Issues and Prospective. Indian Society of Agronomy, IARI, New Delhi, **1** 69–89.

Malik MS, Oraon PR, Kumar A and Singh BK 2018 Agroforestry Systems for Water Stressed Areas. In: Multifunctional Agroforestry, Jaya Publishing House, Delhi, 682–696 pages.

Olsen RK, Schoeneberger MM and Aschmann SG 2000 Ecological foundations of agroforestry, 31–61 pages. In Garrett HE, Rietveld WJ, and Fisher RF (eds.) North American Agroforestry: An Integrated Science and Practice. American Soc. of Agronomy, Madison WI.

Rana SS 2015 Recent Advances in Integrated Farming Systems. Department of Agronomy, College of Agriculture, CSK Himachal Pradesh Krishi Vishvavidyalaya, Palampur, 193 page.

Rietveld WJ and Francis CA 2000 The future of agroforestry in the USA, 387–402 pages. In Garrett HE Rietveld WJ and Fisher RF (eds.) North American Agroforestry: An Integrated Science and Practice. American Soc. of Agronomy, Madison WI.

Sharma LR, Bathi IP, Singh R 1991 Emerging farming systems. Key issues in sustainability. *Indian Journal Agric. Econ.* **46** (3) 422–427.

Singh G 2004 Farming systems options in sustainable management of national resources. In : Proceedings National Symposium on Alternative Farming Systems held at PDCSR, Modipuram, 16–18 September 2004, 80–94 pages.

Tanwar SPS and Tewari JC 2017 Agroforestry and integrated farming system in Draught mitigation and management. Scientific publishers, New Delhi, 181–193 pages.

Tiwari SK 2016 Promising IFS for different regions of India. ICAR–Indian Grassland and Fodder Research Institute, Jhansi (UP).

Tripathi H, Tomar SS, Pandey R, Solanki VS, Singh R, Meena KL, Tomar M and Adhikari DS 2010 Economic feasibility of Integrated Farming System models with respect to productivity and economics. In: 22nd national seminar on "Role of Extension in Integrated Farming Systems for sustainable rural livelihood, 9th – 10th December 2010, Maharashtra, 42–43 pages.

Vandermeer J, Noordwijk M, Anderson J, Ong C and Perfecto I 1998 Global change and multi-species agroecosystems: Concepts and issues. *Agric. Ecosyst. Environ.* **67** 1–22.

Chapter - 18

# Lac Culture based Agroforestry System

## LAC CULTURE

Lac is a resinous exudation from the body of female scale insect. Since Vedic period, it has been in use in India. The English word lac synonyms Lakh in Hindi, which itself is derivative of Sanskrit word Laksh meaning a lakh or hundred thousand. It appears that the Vedic people knew that the lac is obtained from numerous insects and must have also known the biological and commercial aspects of lac industry. With increasing universal environment awareness, the importance of lac has assumed special relevance in the present age, being an eco-friendly, biodegradable and self-sustaining natural material.

### Lac insect

The first scientific name given to it was *Tachardia lacca* following the name of French Missionary Father 'Tachardia'. It was later changed to *Laccifer lacca* Kerr. Lac insect belongs to super family Coccoideae, which includes all scale insects, Phylum Arthropoda, Class Insecta and Order is Hemiptera. Lac is the only natural resin of animal origin secreted by a specialized group of insects called lac insects. Nine (9) genera and 99 species of lac insects have been reported from across the world where 26 species under two genera have been observed in India. Indian lac insect, *Laccifer lacca* (Kerr) is the most important lac insect utilized for cultivation of lac. Kusmi and rangeeni are two strains of *Laccifer lacca* which can be distinguished based on: nutritional requirement, quality of the lac produced, time taken to reach crop maturity and productivity. Other important lac insect species commercially exploited for cultivation of lac are *Kerria chinensis* and *Kerria sharda* (Trivoltine).

### Distribution

Lac is currently produced in India, Myanmar, Thailand, Malaya, Lao and Yuan province of China. India has prime position in relation to lac production. Over 90% of Indian lac produced comes from the states of Bihar, Jharkhand, West Bengal, Madhya Pradesh, Chhattisgarh, Eastern Maharashtra and northern Orissa.

## Host plants

Lac insects thrive on twigs of certain plant species, suck the plant sap, and grow secreting lac resin from their bodies. These plants are called host plants. Lac cultivation requires plants called lac-hosts. More than 400 plants have been reported as lac-hosts all over the world. Kusum (*Schleichera oleosa*), Palash (*Butea monosperma*), Ber (*Zizyphus mauritiana*) and *Ficus* spp. account for about 90 per cent of total lac production in the country as host plants. Besides these, lac cultivation can be commercially carried on about 30 other host-plants. Recently, *Flemingia semialata*, a bushy lac-host has been showing the promise for intensive lac cultivation.

## Lac cultivation on major stand alone host-trees

i. **Kusum** (*Schleichera oleaca)*– It is major host plant for *kusmi* lac cultivation. Mainly two crops of *kusmi* strain (summer season–*jethwi* and winter season–*aghani*) are harvested annually. *Jethwi* crop is cultivated during Jan–Feb to June–July and *aghani* crop is cultivated during June–July to Jan–Feb.

ii. **Palas** (*Butea monosperma)*– It is major host plant for *rangeeni* lac cultivation. Mainly two crops of *rangeeni* strain (summer season–*baisakhi* and rainy season–*katki*) are harvested annually. *Baisakhi* crop is cultivated during Oct.–Nov to June–July and *katki* crop is cultivated during June–July to Oct.–Nov.

iii. **Ber** (*Zizyphus mauritiana)*– It is major host plant for both *kusmi* and *rangeeni* lac cultivation and is employed for winter season *kusmi* crop or summer season *rangeeni* crop.

## Some other Lac Insect Host Plant

- *Acacia catechu* (Khair)
- *Acacia arabica* (Babul)
- *Acacia auriculiformis* (Akashmani)
- *Zizyphus xylopyrus* (Khatber– grown in part of M.P. and U.P.)
- *Shorea talura* (Sal grown in Mysore)
- *Cajanus cajan* (Pigeon-pea or Arhar)
- *Grewia teliaefolia* (Dhaman preferred in Assam)
- *Albizzia lebbek* (Siris/Gulwang)
- *Flemingia macrophylla* (Bholia)
- *Ficus benghalensis* (Bargad)
- *Ficus religiosa* (Peepal)

In India, lac insect is known to have two distinct strains 'kusumi' and 'rangeeni'. Lac cultivation is done by putting lac brood on suitably prepared specific host plants. The lac brood contains gravid females, which are about to lay eggs to give birth to young larvae. After emergence from mother cells, the young larvae settle on fresh twigs of host plants, suck the plant sap and grow to form encrustations.

## Harvesting

Harvesting is the process of collection of ready lac from host trees. It is done by cutting the lac encrusted twigs when crop is mature. It is of two types *i.e.* immature harvesting, the lac is collected before swarming, and lac thus obtained is known as 'ARI LAC'. The drawback of immature harvesting is that lac insects may be damaged at the time of harvesting. In mature harvesting, lac is collected after swarming. The lac obtained is known as mature lac. The mean lac productivity (per tree and per unit area) is 10–15 kg per tree.

## Utilization

- **Lac dye:** Lac dye is a mixture of anthroquinoid derivatives and it is traditionally used to colour wool and silk.
- **Lac wax:** Lac wax is a mixture of higher alcohols, acids and their esters. It is used in polish applied on shoes, floor, automobiles, food and confectionary, lipsticks.
- **Shellac:** Shellac is a natural gum resin, a nature's gift to the mankind and it is non toxic, physiologically harmless and edible resin. It is used in fruit coatings. Jewellers and goldsmiths use lac as a filling material in the hollows in ornaments and also used commonly as sealing wax.

## Lac cultivation in agroforestry

Lac cultivation based agroforestry system, integration of agricultural crops/grasses with host plants by adopting scientific methods plays an important role in making lac cultivation more attractive and a profitable proposition. Intensification and inter-cropping enable the lac farming community to get an additional income, better employment opportunity throughout the year from a particular piece of land besides improving the soil health.

Traditionally, lac is cultivated on three conventional host-plants species, namely, Palash, Ber, and Kusum. Although, several crops can be grown in lac plantation, it is necessary that some choice is exercised, since all crops are not found to perform well under the prevailing plantations. To improve the productivity, profitability as well as for risk management, a number of possible upland agricultural crops and annual and perennial fodder grasses which have been tried are: gora paddy /improved paddy for upland, ragi, green and blackgram, soybean, maize, rhizome, tuber crops, annual grass like Dinanath (*Pennisetum pedicellatem*), Napier grass (*Pennisetum purpurium*) and (*Brachiaria brizantha*). For mixed plantation of Bhalia (*Flemingia macrophylla*) and Galwang (*Albizia lucida*) the best suited intercrops are rhizomes, tuber colocacia etc.

The mixed plantation of Bhalia and Galwang is very suitable for lac cultivation especially for women as all the operations can be carried out from ground level. Bhalia, a shrub and Galwang, a small tree are planted in rows. In one hectare land 3,600 plants of Bhalia and 1200 trees of Galwang are planted. One can start lac cultivation after 4 years of planting. Bhalia is utilised for winter crop and Galwang for summer crop of Kusmi lac.

According to Indian Institute of National Resins and Gums, Ranchi, Dinanath grass (*Pennisetum pedicellatum*) is the best suited fodder grass apart from tuber, rhizome, upland rice, ragi, soybean etc. Integration of various agricultural crops have shown that growing of Dinanath grass + tapioca + turmeric as intercrops leads to 68.8% increase in the gross income over sole lac cultivation on palas; growing sweet potato + ginger + turmeric in mixed plantation of Bhalia and Galwang results in 241.2% increase in gross return. Similarly, profitability of kusum orchard may also be increased.

**Fig.18.1:** *Flemingia semialata* based agroforestry system (IINRG, Ranchi, Jharkhand)

## Economics of Lac cultivation based agroforestry system

Lac cultivation provides sustained & high economic returns, generates employment opportunities and supports lac based rural cottage industries. Out of 25 crore households in India, about one million farm families are engaged in lac cultivation across the country, especially in the eastern and central parts of India. In major lac growing states namely, Chhattisgarh, Jharkhand, Odisha and West Bengal, 25–32 per cent of a family's total farm income comes from cultivation of lac and about one million man-days are generated in the existing lac processing factories alone. Being an export oriented commodity, it contributes towards foreign exchange earning of the country. Globally, India is the leading exporter of lac. Cultivation of lac not only provides livelihood to lac growers, but also helps in conserving the green cover of earth, lac insects and associated biota. India is world leader in lac production with a production of around 17,000 tons of lac per annum (average of 2010–11 to 2014–15). Lac production is confined presently to Eastern India *i.e.* Jharkhand (51%), Madhya Pradesh (15%), Chhattisgarh (44%), and Odisha (4%). Though, classified as minor forest produce on certain considerations, lac ranks as a potential source of revenue for farmers. Growing host-trees for timber and fuel yields revenue in cycles of long years, whereas cultivation of lac on these trees gives a return almost every year, and is therefore preferred by lac growers.

The tribal population mainly depends on rainfed agriculture and forest produce for livelihood. Lac production is an important source of cash flow to the marginal, small and large farmers having very low investment and labour capabilities.

In tribal areas, lac contributes 28 per cent of a farm family's annual income, and is therefore substantive. It also generates employment opportunities for both men and women, particularly in the off-agricultural season.

**Table 18.1:** Income generation by cultivation of lac on major lac-host trees (Rs./ha)

| Particulars | Palas | Ber | Ber | Kusum |
|---|---|---|---|---|
| Lac insect strain | *Rangeeni* | *Rangeeni* | *Kusmi* | *Kusmi* |
| Spacing/planting system | 3.6m × 3.6m | 4m × 4m | 4m × 4m | 12m × 12m |
| Number of trees/plants | 100 trees | 100 trees | 100 trees | 100 trees |
| 1. Gross income | 83000 | 158000 | 258800 | 1122000 |
| 2. Gross cost | 72524 | 117199 | 128814 | 375906 |
| 3. Net income | 10476 | 40801 | 129986 | 746094 |
| 4. Net income per tree | 104.76 | 408.01 | 1299.86 | 7460.94 |
| 5. Net income per hectare | 80770 | 255006 | 812413 | 522266 |
| 6. (Trees per hectare) | (770) | (625) | (625) | (70) |
| 7. BC ratio | **1.144** | **1.348** | **2.003** | **2.985** |

**Table 18.2:** Employment (man-days/ha) generated by cultivation of lac on major lac-host trees.

| Lac-host (No. /ha) | Lac insect strain | Man-days generated | | |
|---|---|---|---|---|
| | | Male | Female | Total |
| *Kusum* (70) | *Kusmi* | 268 | 212 | 480 |
| *Ber* (625) | *Kusmi* | 85 | 73 | 158 |
| *Ber* (625) | *Rangeeni* | 59 | 139 | 98 |
| *Palas* (770) | *Rangeeni* | 57 | 44 | 101 |

*Flemingia semialata:* This leguminous plant is a potential quick growing bush suitable for lac cultivation, as the plantations of *semialata* can be raised within a year and lac crop can be raised in second year of planting.

Management of lac crop on *semialata* is easy as all activities can be carried out from ground itself. The gestation period between two crops is only six months. Intensive lac cultivation on *semialata* is highly remunerative.

**Table 18.3:** Returns from lac cultivation on *Flemingia semialata* (per ha)

| Particulars | *F. semialata* | *F. semialata* |
|---|---|---|
| Spacing/planting system | Paired row system | Single row system |
| Type of lac | *Kusmi* | *Kusmi* |
| Number of plants / hectare | 8000 plants | 10000 plants |
| 1. Gross income | 798000 | 810000 |
| 2. Gross cost | 340925 | 406347 |
| 3. Net income | 457075 | 403653 |
| 4. Net income per plant | 57.13 | 40.37 |
| 5. BC ratio | 2.341 | 1.993 |

**Case Study 1.**

**Table 18.4:** Economic analysis of *F. semialata* based silvi-agri-lac system during first year

| Treatment | Crop | Yield (q/ha) | Rate (Rs./q) | Gross Income Rs./ha | Cost of culti-vation Rs./ha | Net income/ha | B:C Ratio |
|---|---|---|---|---|---|---|---|
| FS + Lac + Cc–LE | Tomato | 20.15 | 4000 | 80,600 | 2,08,000 | 6,85,680 | 3.29 |
| | Arhar | 14.19 | 12000 | 1,70,280 | | | |
| | Lac | 16.07 | 40,000 | 6,42,800<br>=8,93,680 | | | |

| Treatment | Crop | Yield (q/ha) | Rate (Rs./q) | Gross Income Rs./ha | Cost of culti–vation Rs./ha | Net income/ha | B:C Ratio |
|---|---|---|---|---|---|---|---|
| FS + Lac | Lac | 13.55 | 40,000 | 5,42,000 | 1,70,000 | 3,72,000 | 2.18 |
| FM + Lac + CC–LE | Tomato | 22.00 | 4000 | 88,000 | 2,08,000 | 5,36,600 | 2.57 |
| | Arhar | 13.95 | 12,000 | 1,67,400 | | | |
| | Lac | 12.23 | 40,000 | 4,89,200= 7,44,600 | | | |
| FM + Lac | Lac | 8.86 | 40,000 | 3,54,400 | 1,70,000 | 1,84,400 | 1.08 |
| LE | Tomato | 25.50 | 4,000 | 1,02000 | 35,500 | 66,500 | 1.87 |
| CC | Arhar | 14.42 | 12,000 | 1,73,040 | 80,400 | 92,640 | 1.15 |

Where, FS= *F. semialata*, FM= *F. macrophylla*, CC= *Cajanus cajan* and LE= *Lycopersicon esculentum*

*Source:* Berry, N. and Singh, N. 2017

**Table 18.5:** Economic analysis of *F. semialata* based silvi-agri-lac system during Second year

| Treatment | Crop | Yield (q/ha) | Rate (Rs./q) | Gross income Rs./ha | Cost of cultivation Rs./ha | Net income Rs./ha | B:C Ratio |
|---|---|---|---|---|---|---|---|
| FS + Lac + Cc-Le | Tomato | 23.15 | 4,000 | 92,600 + | 1,45,000 | 10,27,560 | 7.08 |
| | Arhar | 15.83 | 12,000 | 1,89,960 + | | | |
| | Lac | 22.30 | 40,000 | 8,90,000= 11,72,560 | | | |
| FS + Lac | Lac | 16.18 | 40,000 | 6,47,200 | 1,05,000 | 5,42,200 | 5.16 |
| FM + Lac + CC–Le | Tomato | 25.00 | 4,000 | 1,00,000 + | 1,45,000 | 6,79,600 | 4.68 |
| | Arhar | 15.15 | 12,000 | 1,81,800 + | | | |
| | Lac | 13.57 | 40,000 | 5,42,800= 8,24,600 | | | |
| FM + Lac | Lac | 10.58 | 40,000 | 4,23,200 | 1,05,000 | 3,18,200 | 3.03 |
| LE | Tomato | 30.25 | 4,000 | 1,21,000 | 30,500 | 90,500 | 2.96 |
| CC | Arhar | 17.07 | 12,000 | 2,04,840 | 65,000 | 1,39,840 | 2.15 |

Where, FS= *F.semialata*, FM= *F.macrophylla*, CC= *Cajanus cajan* and LE= *Lycopersicon esculentum.*

*Source:* Berry, N. and Singh, N. 2017

**Case study 2.** Lac cultivation based agroforestry (Kumar, A. 2017)

Economics of lac cultivation based agroforestry **"Model A"**

Flemingia (7200 plant/ha) + Papaya (1200 plant/ha) + Mango (600 plant/ha). Additionally, between the rows and plant space some vegetable crop *viz.* bhindi, beans, tomato, brinjal etc can be grown to earn additional income.

**Economics**– This way in the first year total expenditure will be Rs. 366700 (Establishment cost Rs. 136500 + annual cost Rs. 230200); gross income will be Rs.1705600 and net income will be Rs.1338900 (gross income Rs. 1705600– expenditure Rs. 366700). The rate and price has calculated as per the year 2013.

Economics of lac cultivation based agroforestry **"Model B"**

Flemingia (7200 plant/ha) + Papaya (900 plant/ha) + Mango (450 plant/ha) + Kusum (450 plant/ha) age 8 years. Additionally, fruit plants and vegetable cultivation will support to add the regular income.

**Economics**– This way in the first year total expenditure will be Rs. 367450 (Establishment cost Rs. 137250 + annual cost Rs. 230200); gross income will be Rs.1565600 and net income will be Rs.1198150 (gross income Rs. 1565600– expenditure Rs. 367450). The rate and price has calculated as per the year 2013.

**Table 18.6:** Economic analysis of Lac cultivation on different hosts

| Host (Strain) | No. of trees | Input 1st yr | Income 1st yr | Input 2nd yr | Income 2nd yr |
|---|---|---|---|---|---|
| Palash (*Rangeeni*) | 100 | 35250.00 | 90150.00 | 15200.00 | 110000.00 |
| Ber (*Rangeeni*) | 100 | 78800.00 | 125000.00 | 50000.00 | 159000.00 |
| Ber (*Kusumi*) | 100 | 120000.00 | 335000.00 | 80000.00 | 450000.00 |
| Kusum (*Kusumi*) | 30 | 170000.00 | 485000.00 | 85000.00 | 625000.00 |
| Semialata (*Kusumi*) | 1000 | 180000.00 | 215000.00 | 70000.00 | 290000.00 |

*Source:* Das & Arvind, 2013

## Conclusion

Lac growers give more importance to regular income from cultivation of lac over the years to one time income from timber or fuel. Thus, lac culture plays a vital role in protection the nation's bio-resources. Lac also provides sustained high economic returns, generates employment opportunities and has potential to lay a strong foundation for lac based rural cottage industries. Lac based agroforestry with high value cash crop that has potential to provide employment and improve socio-economic conditions of the poor and marginal farmers and must be promoted whole heartedly. Farmers are suggested to cultivate lac on *Flemingia semialata* with intercropping of vegetables and fruit plants so that in case of failure, lac crop farmer can get some secured income from the fruit of vegetable crop or *visa-vis.*

## References

Berry N and Singh N 2017 Tropical Forest Research Institute, Jabalpur, Madhya Pradesh

Das R and Kumar A 2013 Lac cultivation and rural livelihood insect host plants of India. In: Kumar, A and Das, R. (edr.) Prospects of scientific lac cultivation in India. Institute of Forest Productivity, 211–231 pages.

Doubling Farmers'Income 2018 Lac Cultivation as an Enterprise. Department of Agriculture, Cooperation and Farmers' Welfare, Ministry of Agriculture & Farmers' Welfare. Volume IX, 50–74 pages.

Kumar A 2017 Lac cultivation based agroforestry. Institute of Forest Productivity, Ranchi–835303 (Jharkhand).

CHAPTER - 19

# Sericulture based Agroforestry System

## SERICULTURE

Sericulture or silk production is defined as breeding and management of silk worms for the commercial production of silk. The word 'Sericulture' has been derived from the Greek word 'Sericos' which means silk, and Culture means cultivation. There are 10 million silkworm rearers and 0.5 million workers related to silk industry in the world (Dewangan, 2013). Asia is the top producer of silk in the world contributing roughly 95% of the total global output. There are 58 countries on the world map of silk; massively produced in China and India, followed by Japan, Brazil and Korea (Nagaraju, 2008). Silk industry is a very important industry in Japan, China, India, Italy, France and Spain. After China, India occupies second position among silk producing countries in the world. Sericulture is an ancient practice in India and plays an important role in livelihood and socio-economic development of rural poor in many areas. The total area under mulberry, on which the best quality mulberry silkworm is reared, is 188000 ha in our country. In India more than 98% of mulberry-silk is produced from four traditional sericultural states *viz.* Karnataka, Andhra Pradesh, West Bengal, Tamil Nadu; and two union territories Ladakh and Jammu & Kashmir.

Sericulture or silk production by rearing silkworms called *Bombyx mori* (mulberry silkworm) has a long and colourful history unknown to most people. This insect is the only living species of family Bombycidae and has been domesticated for so long that it is possible that there are no survivors in the wild any longer. The discovery of silk production from *B. mori* occurred about 2700 BC. The empress found white worms eating the mulberry leaves and formation of shiny cocoons around themselves. Sericulture has been growing in India as an agro-based industry playing a vital role in the improvement of rural economy of India.

### Silkworm

Mulberry silk moth (*Bombyx mori*) is at present time fully domesticated. It no longer exists in wild state and it cannot survive without the human care. The silkworm is the larva or the caterpillar of the moth *Bombyx mori* (popularly known as silk moth) the total life history of the moth from egg to adult take 50 days. The different stages are as follows:

i. Egg 10 days
ii. Larva (4 Stages) 30 days
iii. Pupa (Cocoon) 10 days

(i) **Adult**– The adult silk moth is a creamy white in colour and has a flat body and a wing expanse of about 5 cm. It takes no food and seldom attempts to fly. It lives for only 2–3 days. After mating, the female moth lays 300–500 eggs on leaves of the mulberry tree.

(ii) **Eggs**– The eggs are round and yellowish–white and they become grey as hatching time approaches.

(iii) **Larvae**– The newly hatched larva is about 3 mm long and somewhat black in colour. The larvae grow in size and shed their skin (moult) four times. Each growing stage of the caterpillar consumes more of mulberry leaves.

- ➢ The full grown larva of last stage is about 7 cm long. It has a hump behind the head and a spine–like horn at the tail end.
- ➢ When fully grown, the mature larva stops feeding, climbs on a twig and spins a cocoon.

(iv) **Pupa**– The full grown larva pupates inside the cocoon. In about 10 days, it transforms into a winged adult. The adult moth makes an opening in the cocoon and escapes through it.

## The Cocoon

The cocoon is formed from a secretion from two large silk glands (salivary glands), which extend along the inside of the body and open through a common duct on the lower lip of the mouth parts. The larva moves the head from side to side very rapidly (about 65 times per minute) throwing out the secretion of the silk glands in the form of a thread. The secretion is a clear viscous fluid, which on exposure to the air gets hardened into the fine silk fibre. The filament forming a cocoon is continuous and ranges from 700–1100 m in length.

## Rearing of silkworms

Selected healthy silk moths are allowed to mate for 4 hours. Female moth is then kept in a dark plastic bed. She lays about 400 eggs in 24 hours, the female is then taken out and is crushed. The eggs are hatched in an incubator. The hatched larvae are kept in trays inside a rearing house at a temperature of about 20–25 °C. These are first fed on chopped mulberry leaves. After 4–5 days fresh leaves are provided. As the larvae grow, they are transferred to fresh leaves on clean trays, when fully grown they spin cocoons. After the pupa is transformed into adult, it comes out by cutting the cocoon.

## Reeling silk

The cocoons are cooked in hot water and the silk fibre is unwound from the cocoons. This process is known as reeling. The silk consists of two proteins, the inner core is fibroin and an outer cover of sericin. For reeling silk the cocoons are gathered about 8 days after spinning had begun.

There are following four steps for the completion of the process of reeling:

- The cocoons are first treated by steam or dry heat to kill the insect inside the cocoons. This is necessary to prevent the destruction of the continuous fibre by the emergence of the moth.
- The cocoons are soaked in hot water (95 –97 °C) for 10–15 minutes to soften the gum that binds the silk threads together. This process is called cooking.
- The cooked cocoons are kept in hot water and the loose ends of the thread are caught by hand.
- Threads from several cocoons are wound together on wheels (known as charakhas) to form the reels of raw silk.
- Only about one-half of the silk of each cocoon is reelable, the remainder is used as a silk waste and formed into spun silk.

Raw silk thus obtained is processed through several treatments to give it the final shape.

**Table 19.1:** Hybrid Silkworm Varieties

| Hybrid | Parentage | Importance characters and cocoon yield(kg/100 dfls) |
|---|---|---|
| CSR2 × CSR4 | — | Productive bivoltine hybrid, 65–70 kg /100 dfls |
| CSR16 × CSR17 | — | Productive bivoltine hybrid, 65–70 kg /100 dfls |
| FC1 × FC2 | (CSR6 × CSR26) × (CSR2 × CSR27) | Productive bivoltine double hybrid, 70–75 kg /100 dfls |
| Chamaraja | (CSR50 × CSR51) | Robust bivoltine hybrid, 70–75 kg /100 dfls |
| Jayachamaraja | (CSR50 × CSR52) × (CSR51 × CSR53) | Productive bivoltine double hybrid, 70–75 kg /100 dfls |
| G11 × G19 | — | Double hybrid, 65–70 kg /100 dfls |
| S8 × CSR16 | — | Productive bivoltine hybrid, 70 –75 kg /100 dfls |
| PM × CSR2 | — | Productive multi × bivoltine, 60 kg /100 dfls |
| Cauvery Gold | (MV1 × S8) | Tolerant to silk worm diseases, 75–85 kg /100 dfls |

## Utilization

Silk, being soft, smooth and lustrous, holds a prestigious place among textile fibres and is known as **'Queen of Textiles'**. Raw silk is used for clothing such as shirts, pant, saree, suits, ties, blouses lingerie, pajamas, jackets, handkerchief and other things. Hand spun mulberry silk used for making comforters and sleeping bags. Carpet, furnishing, curtains, draperies, cushion covers, sofa covers and wall hanging.

## Mulberry (*Morus alba*)

Mulberry (*Morus alba*) foliage is the sole food for the mulberry silkworm (*Bombyx mori*) and is grown under varied climatic conditions ranging from temperate to tropical region. Mulberry leaf is a very economic component in sericulture since the quality and quantity of leaf produced per unit area have a direct effect on cocoon quality and quantity. Sericulture broadly comprises of activities such as host plant production, silkworm rearing to produce the silk cocoons, reeling the cocoons for unwinding the silk filament, yarn making, weaving and processing of fabric (Ahmed and Rajan, 2011).

## Distribution

*Morus alba* has a very wide distribution range in Asia and Europe (from Korea to Spain, including China, India); in Africa (North and East Africa) and in South and North America (USA, Argentina, Mexico, Central America, Colombia and Brazil). The origins of most cultivated mulberry varieties are believed to be in the China or Japan area and in the Himalayan foothills. The Cultivation of mulberry plants is called as 'moriculture'. There are about 20 species of mulberry, of which four species, *Morus alba, M. indica, M. serrata and M. latifolia* are commonly cultivated and widely used for silkworm rearing.

## Botanical Description

Mulberry is a fast growing, small–medium sized tree growing up to 10–20 m height. It is generally a short-lived tree. The leaves are usually shiny, dark green and smooth but sometime can be yellowish green. Most leaves are not lobed, but some can be. The juvenile growth is often lobed. Fruits can be white at maturity on a few trees, but are usually dark purple and 3–6 cm long. The fruits ripen from mid spring – late summer (depending on species and cultivar).

## Climatic requirement

Mulberry thrives under various climatic conditions ranging from temperate to tropical located north of the equator between 28° N and 55°N latitude. The ideal range of temperature is from 24 to 28°C. Mulberry grows well in places with an annual rainfall ranging from 600 to 2500 mm.

## Soil

Mulberry flourishes well in soils that are flat, deep, fertile, well drained, loamy to clayey, and porous with good moisture holding capacity. The ideal range of soil pH is 6.2 to 6.8, the optimum being 6.5 to 6.8.

**Table 19.2:** Hybrid Mulberry Varieties

| Hybrid | Parentage | | Importance characters |
|---|---|---|---|
| Victoria–1 | S30 × C776 | CSRTI, Mysore | Tolerant to –leaf spot, powdery mildew, leaf rust |
| G4 | Morus multicaulis × S13 | — | Resistant to leaf rust |
| S13 | Open pollinated hybrid of Kanva–2 | CSRTI, Mysore | Tolerant to –leaf spot, powdery mildew, leaf rust |
| S34 | S30 × C776 | CSRTI, Mysore | Tolerant to –leaf spot, powdery mildew, leaf rust |
| MSG2 | BR4 × S13 | | Quick sprouting after pruning/leaf harvest, High rooting ability (89%) |
| S36 | Mutant of Berhampore local | CSRTI, Mysore | Quick sprouting after pruning/leaf harvest |
| G2 | Morus multicaulis × S34 | — | Quick sprouting after pruning/leaf harvest, High rooting ability (94%) |
| RC1 | Punjab local × Kosen | — | Tolerant to –leaf spot, high yield |
| RC2 | Punjab local × Kosen | — | Quick sprouting after pruning/leaf harvest, High rooting ability (89%) |
| AR12 | S41 (4x) × C776 (2x) | — | Tolerant to –leaf spot, alkaline soil |
| Sahana | K2 × Kosen | — | Leaf yield potential: 25–30 MT/ha/year as inter-crop in coconut plantations |
| BC–259 | Matigare local × Kosen | CSRTI, Berhampore | — |
| Chak Majra | Selection from natural variability | RSRS, Jammu | — |
| China White | Clonal selection | CSRTI, Pampore | — |
| S–146 | Selection from open pollinated hybrids | CSRTI, Berhampore | — |

## Propagation technique

Mulberry is mainly propagated vegetative by cutting, layering or grafting. Propagation by cuttings has been found to be the best method and is the current practice recommended for this purpose. Propagation by seed is only used in tree breeding/tree improvement work.

**Through cutting–** The varieties ideally suited for irrigated conditions are Kanva-2, S-36 and V-1. Six–nine months old branches of about 1.5 cm in diameter should be used for the preparation of cuttings. Cuttings of 15–18 cm in length with three to four healthy buds are used for raising in the nursery or for planting directly in the field.

**Through seed–** Mulberry can be grown from seed and is best sown immediately after fruiting. Cold stratification for 4–16 weeks can improve germination rates. Layering is also reported to work well.

## Spacing

The spacing commonly followed for a rainfed garden is 60 × 60 cm or 90 × 90 cm or 120 × 120 cm. Pits of 35 × 35 × 35 cm are prepared. About 1 kg FYM/pit should be added.

## Irrigation

The plantation should be taken up during the onset of the monsoon to take advantage of the rain. If the rain is not sufficient, the land should be irrigated at regular intervals of 8–14 days depending on the type of soil.

## Fertilization

The total dose of fertilizer to be applied in the first year is 100 : 50 : 50 Kg NPK/ha This is applied in two doses. The first dose is applied when the plantation is about two months old at the rate of 50 : 50 : 50 Kg NPK/ha The second dose is applied after leaf harvesting at the rate of 50 kg N/ha.

## Yield

The crop can yield well for 12 years, after which they are pulled out and fresh planting is done. Yield of mulberry leaves is 30–40 t/ha/year.

## Utilization

Wood is used to extract veeners in the preparation of plywood, paper and pulp industry, fuel wood, sport goods like hockey sticks, toys etc. Mulberry fruit that could be used as a fabric dye, wine, juce, jam etc and leave use silk production.

## Non-mulberry "silks"

1. Tasar silk is produced by certain species of another moth *Antherea royeli.* Their larvae are reared on Arjun (*Termanalia arjuna*), Asan (*Termanalia tomentosa*) and Sal (*Shorea robusta*), chiefly in Jharkhand, Bihar, Madhya Pradesh and West Bengal.

2. Muga silk is obtained from *Antherea assama* whose larvae are reared on Som (*Machilus bombycina*) and Soalu (*Litsea monopetela*) and few other food plants, like Digloti (*Litsaea salicifolia*), Ber (*Zizyphus jujuba*), Champa (*Michelia champaca*), Gamhar (*Gamelina arborea*) etc.
3. Eri silk is produced by the moth *Philosamia ricini*, whose larvae feed on Castor *(Ricinus communis)*, Kesseru (*Heteropanas fragrans*), Borkesseru (*Ailanthus excels*), Gamhar (*Gamelina arborea*) etc. leaves. It is produced in Assam.

## Sericulture in Agroforestry system

There is great potential for Mulberry in agroforestry systems. It's deep-rooting habit and drought tolerance makes it a suitable tree for Alley cropping with grain crops grown in between alleys. Furthermore, the high quality animal fodder that can be produced from the trees make it an excellent choice for agrisilviculture systems although the fodder is generally cut and carried as the plant is not suited to continuous grazing. There is a selection of heavy cropping plants as well plants grown for biomass/animal fodder or sericulture. All of these plants are resistant to all major pest and diseases. Mulberry can be successfully grown as an intercrop (medium mixed tree) between rows of tea/coffee as shade plants. Besides providing shade, a substantial quantity of leaves can be obtained for silkworm rearing and for feeding cattle and goats. Furthermore, the pruned shoots are a good source of firewood. Mulberry can also be grown as an intercrop for cultivation in coconut plantations. A survey has shown that mulberry is intercropped with coconut in the areas of Channapatna, Ramanagaram, Kanakapura and Bangalore in the state of Karnataka. At land management level, sericulture-based silvihorti-pastoral system with mulberry, guava, pineapple (in paired rows) and grasses on bunds was found ideal on the hill slopes (30–45%) with 0. 6–1. 0 m soil depth.

## Economice of Sericulture based Agroforestry system

Sericulture based agroforestry systems on sloping areas combine soil conservation with production function and generate income as well as employment. Sericulture plays a very effective role in the utilization of the natural resources in a most effective manner for socio-economic upliftment with livelihood, employment and income generation (*Malik et al.*, 2008). It is a potential sector of the agriculture to raise economic status of the farming community and also earning foreign revenue (Thapa and Shrestha, 1999).

**Fig. 19.1:** Mulberry based Agroforestry system (JSDI, Ranchi, Jharkhand)

**Table 19.3:** B:C ratio analysis of different sericulture based agroforestry system in India

| Agro-forestry system | B/C ratio | References |
|---|---|---|
| Mulberry (90/150×60 cm) + French bean | 0.93 | Lingaraj, 2008. |
| Mulberry (90/150×60 cm) + field bean | 0.70 | |
| Mulberry (90/150×60 cm) + pea | 1.02 | |
| Mulberry (90/150×60 cm) + carrot | 0.86 | |
| Mulberry (60/120×30 cm) + French bean | 1.10 | |
| Mulberry (60/120×30 cm) + field bean | 0.84 | |
| Mulberry (60/120×30 cm) + pea | 1.52 | |
| Mulberry (60/120×30 cm) + carrot | 1.13 | |
| Sole Mulberry (90/150×60 cm) | 0.57 | |

| Agro-forestry system | B/C ratio | References |
|---|---|---|
| Sole Mulberry (60/120×30 cm) | 0.69 | |
| Sole Mulberry (90×90 cm) farmers practice | 0.39 | |
| Sole Mulberry | – | Bose *et al.*,2009 |
| Mulberry ($K_2O$ 15 kg/ha/year) | 3.27 | |
| Mulberry ($K_2O$ 30 kg/ha/year) | 5.13 | |
| Mulberry ($K_2O$ 45 kg/ha/year) | 4.10 | |
| Mulberry ($K_2O$ 60 kg/ha/year) | 4.69 | |
| **Crop combination: I** | | Misra *et al.*,2009 |
| Sole Mulberry (60×60 cm, 1:1) | 1.04 | |
| Mulberry + Greengram | 1.44 | |
| Mulberry + Cowpea | 1.52 | |
| Mulberry + Toria | 1.27 | |
| Mulberry + Amaranth | 1.40 | |
| Sole Mulberry (90×90 cm, 1:2) | 0.88 | |
| Mulberry + Greengram | 1.50 | |
| Mulberry + Cowpea | 1.59 | |
| Mulberry + Toria | 1.27 | |
| Mulberry + Amaranth | 1.49 | |
| **Crop combination: II** | | |
| Sole Mulberry (60×60 cm, 1:1) | 1.04 | |
| Mulberry + Blackgram | 1.39 | |
| Mulberry + Maize | 1.31 | |
| Mulberry + Pea | 0.85 | |
| Mulberry + Spinach | 1.21 | |
| Sole Mulberry (90×90 cm, 1:2) | 0.88 | |
| Mulberry + Blackgram | 1.32 | |
| Mulberry + Maize | 1.33 | |
| Mulberry + Pea | 0.76 | |
| Mulberry + Spinach | 1.24 | |
| 20 t FYM + 75% Rec. inorganics/ha to mulberry + 12.5 ton FYM /ha to turmeric | 3.42 | Chandrashekar, G. S. 2013 |
| 20 t FYM + 50% Rec. inorganics/ha to mulberry + 12.5 ton FYM /ha to turmeric | 3.40 | |
| 5 t VC + 75% Rec. inorganics/ha to mulberry + 12.5 ton FYM /ha to turmeric | 3.27 | |
| 5 t VC + 50% Rec. inorganics/ha to mulberry + 12.5 ton FYM /ha to turmeric | 3.22 | |
| 20 t FYM + 75% Rec. inorganics/ha to mulberry + 12.5 ton FYM /ha to ginger | 2.50 | |
| 20 t FYM + 50% Rec. inorganics/ha to mulberry + 12.5 ton FYM /ha to ginger | 2.48 | |

| Agro-forestry system | B/C ratio | References |
|---|---|---|
| 5 t VC + 75% Rec. inorganics/ha to mulberry + 12.5 ton FYM /ha to ginger | 2.36 | |
| 5 t VC + 50% Rec. inorganics/ha to mulberry + 12.5 ton FYM /ha to ginger | 2.38 | |
| Paired (20 t FYM/ha + inorganics) | 2.11 | |
| Sole turmeric | 5.44 | |
| Sole ginger | 2.72 | |

## Conclusion

Hence, the above intercropping system in mulberry may be recommended for adoption at the farmers' field for additional income.

## References

Bose PC, Dash BD, Kar R and Bajpai AK 2009 Effect of potassium on growth, yield and nutrient uptake of mulberry (*Morus alba* L.) in eastern ghat region of Orissa. *Journal of Crop and Weed* **5**(1) 34–37.

Chandrashekar GS 2013 Impact of turmeric and ginger intercropping in mulberry on cocoon production under INM. Ph.D. thesis, UAS, Dharwad, 222 page.

Dewangan SK 2013 Livelihood opportunities through sericulture a model of Gharghoda tribal block, Raigarh Dist., *American Journal of Environmental Science* **9**(4) 343–347.

Lingaraj 2008 Mulbery based vegetable intercropping systems under northern transition zone of Karnataka. M.sc. thesis, UAS, Dharwad, 60 page.

Malik MS, Kaushal P and Sah RB 2008 Socioeconomic upliftment of tribal communities in Jharkhand through Agroforestry based farming system. Birsa Agril. University, Ranchi.

Misra AK, Setua GC, Ghosh A, Setua M, Das NK and Bajpai AK 2009 Yield potential and economics of mulberry-based parallel multiple cropping system under irrigated condition. *Journal of Crop and Weed* **5**(1) 48–52.

Nagaraju J 2008 Silk of India, grace and luster. *Biotechnol. News* **3** 4–7.

Thapa RB and Shrestha KB 1999 Silkworm rearing technology. Proceedings of the Workshop at Bhaktapur, Nepal, 21 December 1999.

## CHAPTER - 20

# Apiculture based Agroforestry System

## APICULTURE

Apiculture (Latin word apis = bee, cultura= cultivation) is a branch of agriculture which deals with study of bee keeping and honey production. A location where bees are kept is called an apiary or "bee yard". It is a forest and agro-based industry, which is beyond the ordinary realms of industry in the sense that the humans derive benefits from interaction between two living things like plants and bees without adversely affecting both. Plants, including many crops, prosper and the bees flourish sheltered by humans, giving honey and different other products like bee wax, bee pollen, royal jelly or bee milk, bee-venom, propolis or bee-glue etc. out of which beeswax, propolis and royal jelly are major by-products of beekeeping. Beeswax is used in carpentry, production of candles and cosmetics. Propolis is a substance made by bees from plant resin. It is used for cosmetics, medicine and food. Royal jelly is a nutrient rich substance from beehive. Bee keeping, having its roots in pre–history, can play a very important role in uplifting the livelihoods of especially rural people in India while also benefiting the environment.

### Global scenario

China, Turkey, Argentina, USA and Ukraine are the leading honey producers of the world accounting for 41.71% of the world production. Brazil, India, Tanzania, Ethiopia, Mexico, Kenya and Angola are other major producers of honey. Apart from USA, Canada, Germany and Spain are large producers of honey from the developed countries. In terms for world trade. The largest honey Argentina, China, Mexico, Germany, Hungary are the largest exporters and USA, Germany, Japan, the United Kingdom and France are the largest importers.

Ranking seventh among the honey producing countries, India has been exporting honey since 1991–1992. The quantity exported was around 8,000 tons until 1998, increasing substantially to 15,587 tons in 2009. The major honey–producing states are Punjab, Haryana, Uttar Pradesh, Bihar and West Bengal.

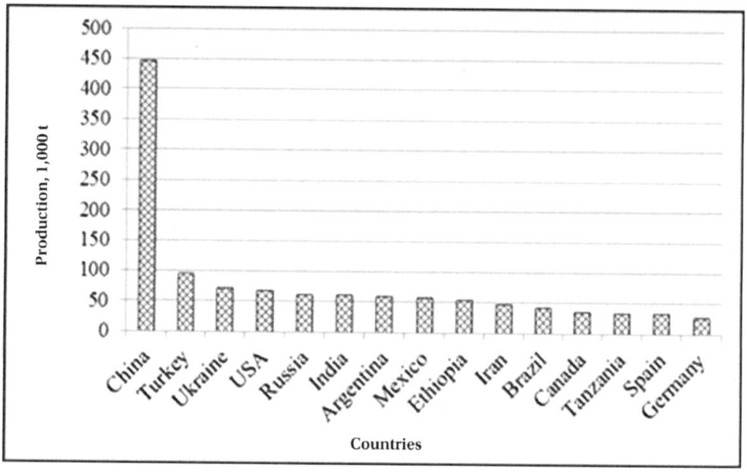


**Fig. 20.1:** Honey production and total beehive number according to some producer countries (*Source:* Anonymous 2013d).

## Honey bee

The modern bee rearing became possible after the discovery of movable frame hives in 1851 by Rerd. L. L. Langshoth. In India, beekeeping was introduced in 1882 in Bengal. Retd. Newton introduced beekeeping to south India in 1911.

Five species of honey bees are found all over the world. There are five important species of honeybees as follows.

- *Apis dorsata*: The rock bee
- *Apis cerana indica*: The Indian honeybee
- *Apis florea* : The little bee
- *Apis mellifera*: The European or Italian bee
- *Melipona irridipennis*: Danner bee, Meliporidae stingless bee.

However *A. cerana* and *A. mellifera* are reared in hives in India for commercial apiculture. *Apis mellifera* (Italian honeybee) was first introduced in 1962 at Nagrota (Kangra) Himachal Pradesh. Presently, this bee species is being used extensively for extraction of honey in India.

## Types of honey

The honey type depends on the dominant kind of nectar sources of the season, which determines the colour, aroma, composition and taste. The composition of honey varies, depending mainly on the source of the nectar(s) from which it is originated and to a lesser extent on certain external factors for *e.g.* climatic conditions and bee keeping practices (White 1975). Of these six types are unifloral honey mainly from the flowers of *Terminalia paniculata* (Honagalu honey), *Syzygium cumini* (Neerilu honey), *Carallia brachiata* (Andamurugilu honey), *Strobilanthus* spp., *Carvia callosa* (Gurige honey), *Sapindus laurifolius* (soapnut or Atalakai honey) and *Gliricidia sepium* (Goppara). Most commonly two kinds are recognized locally *viz.* mixed and soapnut honey. The demand for unifloral honey such as from soapnut is very high compared to mixed honey. It is, therefore, beneficial to plant large number of soapnut trees in the district, especially in the coastal taluks. Even though the quality of honey was identified in three ways like physical, chemical and biological methods, most local people grade the honey by its taste, viscosity, smell and colour.

Forest honey is also named after particular source plant *viz.* Fir Honey, Pine Honey, Lime Tree Honey, Oak Honey etc. Forest honey has higher content of minerals and amino acids, as well as higher molecular weight sugars. It tends to be darker, less sweet, less acidic and resists crystallization when compared to blossom honey (Anonymus 2014a).

**Fig. 20.2:** Apiculture based agroforestry system (BAU, Ranchi, Jharkhand)

## Uses of honey

Honey contains high quantity of fructose which is insulin independent. Honey has been found to contain enzymes (glucose oxidase, catalase), many amino acids (ascorbic acid, flavonoids, phenolic acids and organic acids), carotenoid derivatives and proteins (Gheldof, *et al.*, 2002). It inhibits the growth of micro-organisms like bacteria, fungi and virus. Honey has cleansing action of wounds, stimulates tissue regeneration, reduces inflammation and honey impregnated pads act as non adhesive tissue dressing. It has been used for the treatment of various ophthalmological conditions like lepharitis, keratitis, conjunctivitis, corneal injuries and chemical and thermal burns to eyes (Bansal, *et al.*, 2005).

## Apiculture in Agroforestry system

The bees depend on flowering plants of all life spans such as annuals, biennials and perennials of habits like herbs, shrubs, trees and climbers. Notable examples of some of the annual plants that the bees forage for pollen and nectar are sunflower (*Helianthus annus*) and mustard (*Brassica juncea*). Studies showed yield increase to the tune of 67.7% and 28.08% respectively in mustard, for open pollinated crop and crop enclosed along with honey bees, respectively, compared to control crop enclosed without honey bees (caged without bees) (Harichand and Singh,1995). Nath and Viraktamath (2010) found that *A. dorsata* constituted 96.14% of the pollinators of sunflower followed by *A. florea* (3.35%) and *A. cerana* (0.51%) respectively. Several biennial crops also depend on bees for pollination. Honeybees were found to be predominant pollinators of cauliflower in Haryana. Among honeybees, *Apis dorsata, Apis mellifera, Apis cerana* and *Apis* florea constituted 28.3%, 26.32%, 24.20% and 21.23% respectively (Selvakumar *et al.*, 2003). Perennials or plants that live for several years and producing flowers year after year (For *e.g.* Mango, Banana, Cashew, wild trees and others) are very important in bee-keeping. To maintain beehive populations throughout the year we need a mixture of crop plants including annuals (*e.g.* sunflower, mustard), biennials (*e.g.* radish, sesame) and perennials (*e.g.* mango, soapnut, neem). Such mixed group of species of different habits and life spans, in combination with selected group of wild plants of the region will help in optimizing honey yields almost throughout the year.

Implementation of mixed tree species in apiculture describes that it is the sub-practice of agrosilvipastoral system. In this system, various nectar producing trees and shrubs frequently visited by honeybees are planted on the boundary of the agricultural fields, forest lands and also natural vegetations. Beekeeping is being practiced in tropical orchards, home gardens, plantations, and many agroforestry systems including coconut, coffee, pineapple, and others. Beekeeping is also a relatively low–impact activity that can increase local people's income from native forest or conservation areas.

The genera like *Syzygium, Terminalia, Strobilanthes, Holigarna, Sapindus, Vateria, Lagerstroemia, Phyllanthus, Areca, Cocos, Caryota, Crotalaria, Dalbergia, Pongamia, Pterocarpus, Xylia, Strychnos, Careya* and *Vitex* are some of the very important genera as food sources for honey bees.

**Table 20.1:** List of important pollen /nectar tree spp. sources recommended for promotion of bee farming.

| Sr. No. | Botanical Name | Local Name | English Name | Family |
|---|---|---|---|---|
| 1 | *Acacia auriculiformis* | Bengali babul | Australian Babul | Mimosaceae |
| 2 | *Acacia mangium* | Mangium | Black wattle | Mimosaceae |
| 3 | *Acacia catechu* | Khair | Cutch tree | Mimosaceae |
| 4 | *Alianthus excels* | Arduso | Tree of heaven | Simaroubaceae |
| 5 | *Albizia lebbeck* | Black siras | Womans tongue tree | Mimoseceae |
| 6 | *Albizia procera* | Kilai | White siris | Mimoseceae |
| 7 | *Annona reticulate* | Ramphal | Bullocks heart | Annonaceae |
| 8 | *Annona squamosal* | Sitaphal | Custard apple | Annonaceae |
| 9 | *Anthocephalus cadamba* | Kadamb | Miracle tree | Rubiaceae |
| 10 | *Artocarpus heterophyllous* | Kathal | Jack fruit | Moraceae |
| 11 | *Bauhinia purpurea* | Purplekanchnar | Butterfly tree | Caesalpiniaceae |
| 12 | *Bauhinia racemosa* | Katmauli | Bidi leaf tree | Caesalpiniaceae |
| 13 | *Bombax ceiba* | Lalsemal | Red silk cotton tree | Bombacaceae |
| 14 | *Butea monosperma* | Palash | Flame of the forest | Papilionaceae |
| 15 | *Callistemon lanceolatus* | – | Australian bottle brush | Myrtaceae |
| 16 | *Carica papaya* | Papaya | Papaw tree melon | Cericaceae |
| 17 | *Careya arborea* | Kumbhi | Ceylon Oak | Melastomataceae |
| 18 | *Caryota urence* | Shivjata | Fishtail palm | Palmae |
| 19 | *Cassia fistula* | Amaltash | Indian laburnum | Caesalpiniaceae |
| 20 | *Ceiba pentandra* | Safedshimal | white silk cotton tree | Bombacaceae |
| 21 | *Cinnamomum camphora* | Kapur | Camphor tree | Lauraceae |

| Sr. No. | Botanical Name | Local Name | English Name | Family |
|---|---|---|---|---|
| 22 | *Cinnamomum zeylanicum* | Tejpatra | Cinnamomum tree | Lauraceae |
| 23 | *Dalbergia latifolia* | Shisham | Indian rosewood | Papilionaceae |
| 24 | *Dalbergia sissoo* | Sissoo | Indian redwood | Papilionaceae |
| 25 | *Delonix regia* | Gulmohar | Royal poinoiana | Caesalpiniaceae |
| 26 | *Derris indica* | Karanj | Indian beech tree | Papilionaceae |
| 27 | *Dillenia indica* | Karmal | Elephant apple | Dillaniaceae |
| 28 | *Diospyros melanoxylaon* | Tendu | Ebony | Ebenaceae |
| 29 | *Emblica officinalis* | Amla | Indian goose-berry | Euphorbiaceae |
| 30 | *Eucalyptus tereticornis* | Nilgiri | Eucalyptus | Myrtaceae |
| 31 | *Ficus benghalensis* | Bargad | Banyan tree | Moraceae |
| 32 | *Ficus religiosa* | Pipal | Sacred ficus | Moraceae |
| 33 | *Gmelina arborea* | Gamhar, sevan | White teak | Verbenaceae |
| 34 | *Grewia asiatica* | Phalsa | — | Malvaceae |
| 35 | *Jacaranda mimosifolia* | Neel mohar | Jacaranda | Bignoniaceae |
| 36 | *Lagerstromia speciasa* | Taman | Pride of India | Lythraceae |
| 37 | *Leucaena leucocephala* | Subabul | Ipilipil | Mimosaceae |
| 38 | *Litchi chinensis* | Lichi | Litchi | Sepindaceae |
| 39 | *Madhuca indica* | Mahuda | Butter tree | Sapotaceae |
| 40 | *Mangifera indica* | Aam | Mango tree | Anacardiaceae |
| 41 | *Manilkara zapota* | Chikku | Sapota | Sapotaceae |
| 42 | *Michelia champaca* | Son champa | Golden champa | Magnoliaceae |
| 43 | *Mimosops elengi* | Bakul | Indian modlar | Sapotaceae |
| 44 | *Mitragyna parvifolia* | Kalam | Etagalu mara | Rubiaceae |
| 45 | *Moringa oleifera* | Sahjan, Drumstick | House radish tree | Moringaceae |

| Sr. No. | Botanical Name | Local Name | English Name | Family |
|---|---|---|---|---|
| 46 | *Morus alba* | Shetur | Mulberry | Moreceae |
| 47 | *Nerium indicum* | Lalkaren | Dleender | Apocynaceae |
| 48 | *Pithocaliabium dulce* | Jangalijalebi | Manila tamarind | Mimosaceae |
| 49 | *Psidium guajava* | Amrud | Guava | Myrtaceae |
| 50 | *Punica granatum* | Anar | Pomegranate | Punicaceae |
| 51 | *Semenea saman* | Ratosarasado | Raintree | Mimosaceae |
| 52 | *Santalum album* | Chandan | Sandal wood | Santalaceae |
| 53 | *Spathodea campanulata* | Fountain tree | African tuliptree | Bignoniaceae |
| 54 | *Schleichera oleosa* | Kusum | Lac tree | Sepindaceae |
| 55 | *Shorea robusta* | Sal | Resin tree | Dipterocarpaceae |
| 56 | *Syzygium cumini* | Jamun | Black plum | Myrtaceae |
| 57 | *Temarindus indica* | Emli | Tamarind | Caesalpiniaceae |
| 58 | *Tecoma stans* | Tecoma | Yellow elder | Bignoniaceae |
| 59 | *Tectona grandis* | Sag | Teak | Vervenaceae |
| 60 | *Terminalia cetappa* | Deshibadam | Indian almond | Combretaceae |
| 61 | *Terminalia chebula* | Harde | Chebulicmyrobalan | Combretaceae |
| 62 | *Terminalia bellirica* | Bahrda | Bellericmyrobalan | Combretaceae |
| 63 | *Vitex negundo* | Nagod | Chwte tree | Verbenaceae |
| 64 | *Zizyphus mauritiana* | Ber | Indian cherry | Rhamnaceae |

## Economics of Apiculture based Agroforestry system

Apiculture is a technical procedure and an economic activity done in most of the regions of India. Apiculture makes a great contribution to provide food and income security for the people living in rural areas. Beekeeping can be practiced successfully on a part-time basis, and yields a wide array of high-value products that can increase a farmer's income by 40–60%. Agroforestry, growing crops with trees and shrubs provide nectar for bees. The improved pollination by the bees enhances production of seeds or fruits of these plants as well.

**Table 20.2:** Increase in yield by honey bee (*Apis cerana*) pollination of fruit tree and crop productivity

| Tree | Increase in fruit set (%) | Increase in fruit weight (%) | References |
|---|---|---|---|
| Apple | 10 | 33 | Verma and Dulta, 1998 |
| Peach | 22 | 44 | Partap *et al.*, 2000 |
| Plum | 133 | 39 | Partap *et al.*, 2000 |
| Citrus | 24 | 35 | Partap *et al.*, 2000 |
| | **Agriculture crops** | | |
| **Crop** | **Percent (%) yield increase** | | **References** |
| Mustard | 44 | | Honey Bee Farming Information Guide, Agri Farming, 2018. |
| Sunflower | 32–45 | | |
| Cotton | 17–20 | | |
| Lucerne | 110 | | |
| Onion | 90 | | |

## Conclusion

Apiculture is the important agroforestry system in the world, including India, and is also perhaps the oldest system. As a general approach, when designing agroforestry systems, the use of a range of tree and shrubs species flowering at different times can provide better forage for bees, therefore enhancing both honey production, tree products and annual crop production. Beekeeping can be practiced successfully on a part-time basis, and yields a wide array of high-value products that can increase a farmer's income by 40–60 percent.

## Reference

Anonymous 2013d Honey: World Production, Top Exporters, Top Importers, and United States Imports by Country. Phong Son Honey Company. Available: http://honeyvietnam.com /en/component/content/article/11–news/16–honey-world-production-top-exporters-top-importers-and-united-states-imports-by-country.html

Anonymous 2014a Honeydew honey or forest honey. Available: http://www.honey traveler .com/single–flower–honey/ honeydew–or–forest–honeys/

Bansal V, Medhi B and Pandhi P 2005 Honey–A remedy rediscovered and its therapeutic utility. *Kathmandu University Medical Journal* **3**(11) 305–309.

Gheldof N, Wang XH and Engeseth NJ 2002 Identification and quantification of antioxidant components of honeys from various floral sources. *J Agric Food Chem.* **50** 5870–5877.

Honey Bee Farming Information Guide, 2018. *Agri. Farming,* 1–9 pages.

Nath S and Viraktamath S 2010 Pollinator fauna of sunflower and their relative abundance. *Karnataka J. Agric. Sci.* **23**(3) 517–518.

Partap U, Shukla AN, and Verma LR 2000 Comparative attractiveness of broad-leaf mustard, cauliflower and radish to Apis cerana in Kathmandu Valley of Nepal. In: Proceedings of the 4th Asian Apicultural Association International Conference, Kathmandu, 203–205 pages.

Ramachandra TV, Subash Chandran MD, Joshi NV and Balachandran C 2012 Beekeeping: Sustainable Livelihood Option in Uttarakhand, Central Western Ghats. ENVIS Technical Report: 49.

Selvakumar P, Sinha SN, Pandita VK, and Srivastava RM, 2003 Foraging behaviour of honeybee on parental lines of hybrid cauliflower push hybrid-2. 38$^{th}$ Congress Apimondia, Ljubljana, Slovenia.

White JW 1975 Composition of honey. In Crane, E. (ed) Honey: A Comprehensive Survey Heinemann ed. London, 157–206 pages.

Wilkinson K and Elevitch C 1999 Posted in The Overstory eJournal (/overstory-back-issues)/ Overstory #40 – Bees and Agroforestry.